Math and Magic in Wonderland

Lilac Mohr

For Linnea, Elowen, Hallden, and Linden.

To four future mathematicians
Who, like very short magicians,
Have somehow gained admission
(Though I surely gave permission)
To the center of my heart.

Contents

Letter from the Author

Dear Reader,

Have you ever wanted to learn something only to be told that you were too young to understand it? That's exactly what happened to me when I was in the second grade.

At the time, my hero was Maria Mitchell, a female astronomer who is best known for discovering a comet. I remember the day that I read in Maria's biography that she was able to do long division at the age of eight. I slammed the book shut and ran excitedly to my teacher's desk to ask her to teach me long division, too. The teacher rejected my enthusiastic request and explained that I would have to wait until the fourth grade. She also reminded me not to talk during 'silent reading time'.

I felt dejected but not defeated and decided that if my teacher refused to teach me long division, I would have to simply teach myself. It was not easy (since I didn't even know multiplication at the time), but with determination, I pressed through and mastered the skill on my own.

Why am I telling you this story? Now, years later, I still recall my teacher's rejection as the moment I realized that curiosity should not have age restrictions. I wrote this book with the intention of making the magic of math accessible to everyone regardless of age, gender, or background. All you have to bring on this journey is a love of learning.

I invite you now to join two exceptional girls on a grand adventure to Wonderland. Keep a piece of paper handy, so you can play along with our heroines. Each problem has a point value, and if you keep track of your score, at the end of the book you can see if you have what it takes to be crowned Math Royalty. You can also find additional resources for each chapter on my blog: *http://learnersinbloom.blogspot.com*. I hope you enjoy this jaunt into the world of Math and Magic!

Sincerely,
Lilac Mohr

"My Princess," he said tenderly, "two great powers are on our side: the power of Love and the power of Arithmetic. Those two are stronger than anything else in the world." (E. Nesbit)

1. Mrs. Magpie's Manual

Every heroine needs a trusty sword, thought Lulu as she ran her fingers across a toy sword, dreaming of its heroic exploits. The sword was made of floppy craft foam, rendering it a rather unconvincing weapon. To Lulu, however, a flaccid toy sword was as good as any other. "Want to play King Arthur?" she asked her sister.

There was no answer from Elizabeth, who was lying in bed, deeply engrossed in her book. Lulu, sword in hand and stomach flat on the carpet, crept toward the foot of her twin sister's bed. She lay there quietly for a moment, listening. Elizabeth turned a page and sighed. *Good*, thought Lulu, *I haven't been detected.*

Lulu, still lying on the floor, rolled sideways and slowly lifted the sword above her head. She then completed her ascent, carefully and deliberately rising- first on her knees, and then in a single flowing motion onto her feet. "I am the Lady of the Lake," she spoke in a ghostly voice, "here to present you with Excalibur." She gave a low curtsy and lowered the sword onto her sister's bed. The brilliance of this performance had been completely lost on Elizabeth who had not glanced up from her book.

"I'm doing some serious research here," said Elizabeth. "Did you know that there is a type of marsupial mouse that can breathe through its skin until its lungs are developed? It's the only mammal that can do so." She finally lifted her gaze away from the book and couldn't hide the look of surprise that covered her face as she saw Lulu, attired in a sequin skirt and tiara, bowing before her bed.

"Aren't we too old for these types of games?" Elizabeth laughed. Even though she and her sister were born only eight minutes apart, Elizabeth often felt like she had to be the more mature twin.

"Never!" Lulu replied emphatically. Suddenly a new idea flew into her head. "I know," she said, sword waving. "We can be Celts, fighting Julius Caesar. Maybe Mother will let us paint our faces blue, again."

"What? No way!" Elizabeth shook her head and tried to suppress a smile as she lifted the book in front of her face. "Besides, last time the paint didn't wash off completely," she added. "We looked like a pair of Smurfs for a week, remember?"

"I don't suppose you'd want to be Odysseus?" asked Lulu, mischievously, eyeing a pink jump rope on the floor. "I can tie you to the bed so you will be immune to the lure of the enchanting sirens, played by me, of course." She was well aware of the answer that would follow, so this was more of an attempt at comedic appeal.

"Not a chance," Elizabeth snickered, her face still eclipsed by the book.

"You leave me no choice," Lulu said gravely, "but to slay the Jabberwock by myself."

"I think you're getting your stories all mixed up-" Elizabeth began. This time, it was Lulu who was not paying attention. She had already sprung up on her feet, sword in hand.

Lulu waved the flopping sword as she skillfully stepped around the toys, books, and clothes that were strewn around the floor.

"One, two! One, two! And through and through
The vorpal blade went snicker-snack!
He left it dead, and with its head
He went galumphing back."

She picked up a stuffed bear and galloped back to her bed with it. Elizabeth peeked up from her book just long enough to roll her eyes.

"And hast thou slain the Jabberwock?
Come to my arms, my beamish boy!
O frabjous day! Callooh! Callay!"
He chortled in his joy."[1]

Lulu was now on top of her bed dancing triumphantly. "Why aren't the girls ever the ones slaying monsters?" she wondered aloud.

"Because we are too smart for that," her sister replied sensibly. "Can you read or do something quietly for a bit. We'll work on a play to perform after dinner. Maybe Little Women." There was silence as Lulu pondered the offer.

"You can be Jo," Elizabeth added enticingly.

"Deal," agreed Lulu. She sat down on her bed, but it was hard to relax, her head still racing with dramatic play ideas. The only logical course of action was to recite the digits of Pi. Lulu had done this so often as a tactic for calming down, that the numbers simply came out instinctively. It was like singing a song she had heard a million times before. "Three point one, four, one, five, nine, two, six, five, three, five, eight, nine," Lulu recited rhythmically. By the time she had spouted out the first one-hundred digits, Lulu's mind was serene and ready to focus. *But focus on what?*

Lulu decided that writing the script for their play did not only count as a quiet activity but would ensure that Elizabeth kept her word. She tip-toed to the windowsill to retrieve the colorful pens her sister had left there from that morning's nature sketching session.

Suddenly, the dark silhouette of a bird filled the window-frame. A large black-blue magpie had landed on the sill of the open window. Lulu found herself face to face with the elegant creature. The magpie cocked its head and stared at Lulu with its shining eyes. Before Lulu had a chance to exhale, the magpie picked up a pen in its claws and flew away on silent wings.

"Did you see that?" Lulu yelled out in surprise.

"What now?" asked Elizabeth, irritated.

"There was a magpie at the window just now and it looked at me with a sort of twinkle in its eye like it was sending me a message and then

it stole a pen and we need to find it right away."
Lulu, who often spoke in run-on sentences when
excited, was out of breath. She grabbed her
backpack and began filling it with essential
adventuring supplies – a calculator, ruler,
compass, and protractor. "Get your shoes on.
Hurry!" she yelled to her sister.

"Nice try," said Elizabeth flatly, "but I
would have heard it. Owls can fly silently, not
magpies."

"Please believe me. This is important,"
Lulu pleaded as she shoved a stack of scratch-
paper and a handful of pencils into her bag.
Then she spotted it - a feather lying on the
windowsill. Its edges were midnight black, and
its center was creamy white. Lulu grabbed the
feather and dropped it on the page of the open
book in her sister's lap. "Please believe me," she
repeated. "Come on."

Elizabeth studied the black-and-white
feather but did not make a motion to stand up.
"A pen can be replaced," she finally said. "You
can even turn this feather into a quill pen if you
want."

But the pitiful look on her sister's face soon changed Elizabeth's mind. She slipped on her shoes. "If you're after a wild-goose chase, Don Quixote, I'll be your Sancho Panza."

"A wild-magpie chase, you mean," laughed Lulu, "and I promise not to slay any windmills."

Lulu slung the backpack across her shoulders as she jetted downstairs and out the back door. Elizabeth followed.

"What's the plan?" Elizabeth asked, squinting as her eyes adjusted to the bright sunlight.

Lulu, who rarely made plans and seldom followed the ones she made, looked around. To her great delight, she spotted the magpie sitting on a wooden fencepost in the park across the street. Without pausing to think, Lulu placed the magpie feather in her pocket and dashed toward the bird.

Elizabeth, driven more by obligation than enthusiasm, jogged behind her sister. She was astonished to see that instead of flying off at the sight of a girl rushing straight at it, the magpie waited until Lulu stopped right in front of the

11

fence before flying off. Lulu, of course, was right at its heels, darting around park benches and leaping over shrubs.

The magpie alighted on a tall maple. Lulu was staring, mouth agape, at the crimson-clad tree-boughs when Elizabeth reached her. "I caught a glimpse of the pen it stole," said Lulu. "It was the silver glitter gel pen - my favorite!"

"So what?"

"If that isn't a sign, then I don't know what is. I need to get that pen back," Lulu continued excitedly.

"Are you going to climb the tree, Tarzan? What's your plan?" Elizabeth wondered.

Lulu was already calculating which limbs would best support her body weight and did not reply.

"What's your plan?" Elizabeth asked again.

"A man, a plan, a canal, Panama!" responded Lulu. She grabbed a branch and began climbing.

Elizabeth sighed and shook her head. She watched as her sister adroitly moved from bough to bough. Soon, however, Elizabeth's artistic eyes caught the play of light and shadow among the leaves of the tree as they trembled in the autumn breeze. Before long, she had forgotten her sister and was completely engrossed in the leaves' dance. She remembered a poem by W. H. Davies:

> *What is this life if, full of care,*
> *We have no time to stand and stare?*[2]

Elizabeth took a deep breath. There was a subtle scent of smoke in the air. She was trying to recall the poem about autumn bonfires by Robert Louis Stevenson when a pen fell sharply on her forehead, narrowly missing her eye and putting an abrupt end to Elizabeth's reverie. A pair of tan legs dangled from the branch above her.

"The magpie dropped the pen," Lulu called out, jumping down onto the grass, "and you'll never believe what I found!"

Elizabeth rubbed the sore spot on her forehead and through blurry tear-filled eyes saw that Lulu was holding a thick carmine-colored book in her hands.

"There was a hole in the tree, about half-way up. And this was inside," Lulu breathlessly explained.

"Oh no, are you hurt?" she asked, seeing her sister's face.

"I'm fine. Let's see what you found," replied Elizabeth bravely, not wanting to mar her sister's enthusiasm.

The twins leaned over the book. An elegant letter "M" was embossed in gold on its cover.

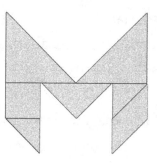

"*Mrs. Magpie's Manual of Magic for Mathematical Minds,*" Lulu read aloud. Her heart fluttered. *Math and Magic? It's as if this book was meant for me.*

"Actually, most authors avoid alliteration nowadays," chimed in Elizabeth.

Lulu began laughing. "Authors actually avoid alliteration? Always?"

"Most of the time," Elizabeth responded earnestly, oblivious to her own accidental joke. "I saw it in a list of top ten mistakes new authors make."

Lulu, feeling there was no time to waste, eagerly opened *Mrs. Magpie's Manual of Magic for Mathematical Minds* to a random page. It was completely blank. She tried a different location and it, too, was blank. She flipped frantically through the book, but the result was always the same – blank. A crestfallen look washed across Lulu's face.

"Maybe it's an empty diary or a geocache log that nobody has found yet," speculated Elizabeth who was looking over her sister's shoulder.

Lulu was not a girl who accepted defeat so readily. *I'll approach this puzzle using logic,* she told herself. *If the book is magical, then maybe uttering a magic spell would reveal the words.*

"Where to start? Where to start?" she said aloud, quickly running through the list of famous magic words, from "abracadabra" to "open sesame", in her head.

"Starting at the beginning is usually a good idea," suggested always-sensible Elizabeth.

Lulu turned to the first page of the book. *It wasn't empty!* Her hands began to tremble in excitement. They trembled so much, in fact, that reading the elaborate script was nearly impossible. There was only one thing to do. "Three point one, four, one, five, nine-" Lulu began.

"Why can't you just count to ten to calm down like everyone else?" asked Elizabeth. "Give me that," she reached for the book.

"I prefer Pi," retorted Lulu, her hands clasping the book even tighter. In a clear voice, she read aloud:

Life is too short for long introductions
With run-on sentences full of conjunctions.
So right to the problems I'll dare to venture.
Solve them to go on a grand adventure.
Take a deep breath, and don't get vexed.
Each solution will unlock the one that comes next.

"Aha!" said Elizabeth, releasing her grip on the book, "That explains it."

Lulu could only stare at the page dreamily. "A grand adventure," she sighed.

"Keep reading," pressed Elizabeth. She was skeptical that this book held any magical properties, but the draw of a good puzzle was irresistible.

To reveal the problem on the following page,
Your first task is to determine my age.
In your world, a magpie's lifespan is twenty-three,
But things are quite different in this one, you see.
The number of years I have seen since my birth
Are a century more than twice a lifespan on Earth.

Play Along: (2 pts.) Help Lulu and Elizabeth solve the first clue. How old is Mrs. Magpie? Keep reading for the answer.

"A century is 100 years, like 100 cents in a dollar and 100 centimeters in a meter," Lulu reasoned aloud.

"Right," agreed Elizabeth. She was thinking about how a centipede doesn't really have one hundred legs but didn't let her mind wander for long. "Mrs. Magpie must be a centenarian. But what does *'twice a lifespan on Earth'* mean?"

"Maybe Mrs. Magpie lived and died on Earth before being transported to another world," conjectured Lulu. "Or maybe," she continued before her sister could interrupt, "Mrs. Magpie lives on a planet with shorter years than Earth. Mercury, for example, takes only 88 days to orbit the sun, compared to Earth's 365 days. That would mean that approximately four years pass in Mrs. Magpie's world for every year on Earth. Hand me the calcula-"

"Hold your horses, Don Quixote," said Elizabeth, relieved at finally getting a chance to be heard. "You promised not to jump to conclusions. The clue didn't even mention Mercury. Let's think through this logically."

"I only promised not to mistake windmills for giants," corrected Lulu, "but this is totally different."

Elizabeth furrowed her brow, deep in thought. "I think we better use the information we were given," she finally said. "The life expectancy of a magpie is 23 years. So double that number-"

"46"

"-and add 100."

"Mrs. Magpie, you are one hundred forty-six years old," Lulu shouted at the top of her voice, looking up into the tree.

"What are you doing?" asked Elizabeth, looking quickly around to make sure that she was the only witness to her sister's shenanigans.

"I'm telling Mrs. Magpie our answer, of course."

"Do you really think the bird we saw earlier wrote this book?" chuckled Elizabeth. "I don't think it's even in the tree anymore. How about writing the answer down like a normal person?"

Lulu, who would have been more insulted had she been *called* 'normal', grabbed the silver glitter pen and quickly wrote down "146-years-old".

At once, the characters Lulu had just written began to glow. Lulu glanced over at Elizabeth, whose large eyes shone with a mixture of surprise, fear, and delight. *Good*, thought Lulu, *she can see it too*. Without hesitation, Lulu quickly flipped to the following page. It was no longer empty.

"That was almost too easy," Elizabeth commented. She suddenly caught a glimpse of brown and white fur on the other side of the tree-trunk. "Was it a cat I saw?" Elizabeth peered behind the tree. "No, it was a squirrel," she answered herself, then hesitated. "But it can't possibly be!"

Lulu looked up at the cute critter that was now sitting on the branch above them. Its body was dark gray, almost black, and its tail was white and fluffy. The two dark eyes on its chestnut brown head seemed to be staring at her with curiosity. "I'm pretty sure that's a squirrel," Lulu concluded.

"Of course it is. But this one is a Kaibab squirrel," Elizabeth explained. "The only place they live is around the Grand Canyon. Something highly unusual is going on here."

"So let me get this straight," said Lulu. "We find a magical book, written by an anthropomorphic magpie, which will take us on a grand adventure, presumably to a different dimension... and you think a squirrel is 'highly unusual'?" She shrugged and began reading the next clue.

Play Along

At the end of each chapter, you will get a chance to play along with Lulu and Elizabeth. Be sure to always work on a separate piece of paper, not directly in the book. Solutions are found in Appendix A.

1. (1 pt.) Assuming authors actually allow alliteration; make up your own silly sentence using words that begin with the first letter of your name.

2. (3-18 pts.) Lulu recites digits of Pi to calm down. Pi is a special number that represents the ratio between the circumference of a circle (the distance around it), and its diameter (the distance across the circle, through its center). The decimals of Pi go on forever without any pattern. You'll get a chance to explore Pi further in the fourth chapter. Just for fun, see how many digits

of Pi (after the decimal point) you can memorize. Here are the first thirty:

Pi = 3.141592653589793238462643383279

Give yourself 3 points for every 5 digits you memorize.

3. (5-10 pts.) "A man, a plan, a canal, Panama!" is a famous palindrome, a sentence (or number) that reads the same both forwards and backward. There are two additional sentences that are palindromes hidden in this chapter. Give yourself five points for each one you find. Hint: One palindrome is spoken by Lulu and the other by Elizabeth.

4. It's also fun to play with numeric palindromes (numbers that read the same forwards and backward):

   a. (1 pt.) How many 2-digit palindromes (like 11 and 22) exist?

b. (2 pts.) How many 3-digit palindromes (like 101, 111, and 121) exist? Hint: You don't have to write them all down - just look for a pattern.

5. If Mrs. Magpie is 146 years old in Earth years, what is her age in:

a. (2 pts.) Mercury years? (Four Mercury years pass for every Earth year.)

b. (2 pts.) Mars years? (Mars only completes half its orbit around the sun in one Earth year.)

2. Magic Square

One ancient game of skill will soon reveal another.
First place the seven Tans to form a turtle brother.
Then write how many fewer rights
There are than all the others.

The two girls stared at each other in a rare moment of silence. "Any ideas?" Lulu finally asked.

"*Tans* is capitalized," noted Elizabeth. "Maybe it stands for something."

Lulu's eyes twinkled. "This book is a manual of magic for *mathematical* minds, right? Tan must stand for a math term like 'tangent'. A line is tangent to a circle if it touches it at just one point."

"So we need to draw seven lines tangent to the circle to form the shape of a turtle!" Elizabeth already had a ruler and pencil in hand.

"You're going off on a tangent," a small voice seemed to come out of nowhere.

Lulu froze in surprise. "Did you hear that?" she asked.

"What did you say?" Elizabeth was busy planning her turtle.

Play Along: (2 pts.) Trace a circle. Can you make it look like a turtle (add 1 head, 4 legs, and 1 tail) using only seven straight lines, all tangent to the circle? Keep reading for the answer.

"If I draw the lines so the circle is inscribed in two triangles which point in opposite directions, I think I can make a turtle using only six lines," Elizabeth thought aloud, "but it wouldn't look that much like a turtle."

"That shape is called a hexagram," chimed in Lulu, who was listening, "and I don't think it's right. We're going off on a tangent."

"Very funny," said Elizabeth flatly. She rotated the ruler around, puzzling over where the seventh line might go.

That voice was right - we're going down the wrong path, decided Lulu. She flipped to the cover of *Mrs. Magpie's Manual*, hoping for inspiration. "Look at this letter M," she cried.

Elizabeth was annoyed by the distraction. "What does that letter have to do with the riddle?" she asked impatiently.

"Look closely," insisted Lulu. "The letter M is made up of smaller geometric shapes."

Elizabeth's scowl vanished. "Tangrams!" She hid the ruler behind her back, suddenly feeling foolish. This act did not escape her sister's eyes. *Which of us is Don Quixote now?* Lulu often lacked the tact to keep her witty (and sometimes tactless) remarks to herself. On this occasion, however, she was intent on solving the puzzle.

"Seven Tans, an ancient game of skill; this has to be it," Lulu shouted gleefully. She traced the shapes with her fingertips. As her fingers pressed down on one of the small triangles, the seven pieces tumbled into the palm of Lulu's hand.

Lulu handed the Tangram shapes to her sister, who had been fidgeting with anticipation. It wasn't long before the form of a turtle filled the blank page of the open book.

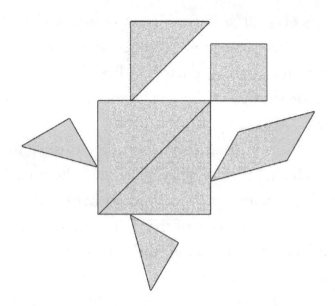

"It's a sea turtle," declared Elizabeth. "See the flippers?"

"More like a Mock Turtle if you ask me," joked Lulu. "I've never seen a turtle whose flippers are made up of three triangles and a diamond."

"Parallelogram," corrected Elizabeth, feelings hurt. "If you say diamond, someone might think you are talking about a rhombus. But this shape's sides are not all the same length, so it can't be a rhombus."

"What's in a name? That which we call a rose by any other name would smell as sweet.₃" Lulu posed dramatically with a hand over her heart.

"O.K., Shakespeare, you know very well that math doesn't work like that. A rhombus by any other name…" Elizabeth's voice trailed off for a second.

"…would be the wrong bus," a small voice sounded from above.

"The wrong bus?" Elizabeth asked aloud as she looked around in all directions.

"Great pun- rhombus, wrong bus," Lulu chuckled.

"There was a voice," Elizabeth said meekly. She paused, thinking about how to explain what she heard.

Lulu understood at once. "The same one I heard before! It told us we were going off on a tangent, and now apparently we're taking the wrong bus."

Elizabeth thought for a moment. "Maybe we are getting off track. Let's focus on the puzzle," she said quietly.

"*First place the seven Tans to form a turtle brother*," read Lulu. "We did that part."

"*Then write how many fewer rights there are than all the others*," she continued reading.

Elizabeth shrugged. "Write and right are homophones," she offered.

"Right," agreed Lulu with a giggle. Then she had a new thought. "Five of the Tangram pieces are right triangles... umm I mean right isosceles triangles, since we're being precise here," Lulu corrected herself mockingly before Elizabeth even had a chance to open her mouth. "A right isosceles triangle by any other name... would be obtuse!" She laughed loudly.

Elizabeth smiled and shook her head. "Cute," she said

"You mean acute!" came the small voice. This time, both girls heard it and looked up at the tree. The squirrel was sitting on the branch above their heads staring. As if in answer to the question in both girls' minds, it chattered in a ⸻ squirrel-like manner (more like a red squirrel ⸻ Kaibab, Elizabeth would later recall), and ⸻ up the tree.

"This day is getting stranger and stranger," Elizabeth said, putting words to what both girls were thinking.

"Right, obtuse, acute," said Lulu aloud. "Get my angle?"

"Angles! We need to count the angles," said Elizabeth. "And stop with the puns. Please," she pleaded.

The twins traced the outline of the Tangram turtle in the book and then set about counting the right, acute, and obtuse angles on each of the Tans.

Play Along: (5 pts.) What number should Lulu and Elizabeth write on the turtle's shell? The seven shapes include five right triangles, a square, and a parallelogram. In those Tangram pieces, how many fewer right angles are there than the other types of angles? Keep reading for the answer.

Elizabeth wrote the answer in the center of the turtle's shell. As soon as she did so, a square, divided into three rows and three columns, appeared on the turtle's back, along with a new set of instructions beneath it.

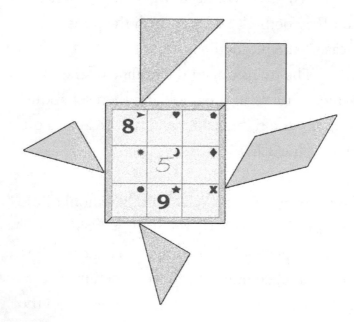

"I think I know what this is," said Elizabeth "but we better read the instructions to ~ertain." Lulu read them aloud:

"Sum and one don't rhyme," said Elizabeth.

"Who cares?" said Lulu. "Didn't you hear the clue? We have to solve the magic square."

"That's what I thought it was. All the rows, columns, and diagonals are supposed to add up to the same "magic constant". But what is that sum?

"I know how to figure that out," Lulu said, excitedly. "There are nine cells, right?"

"Yes," answered Elizabeth.

"And each cell gets filled with a number from one to nine, no duplicates. Adding up the numbers in each of the three rows will give you the same total, correct?"

"Sure. What are you getting at?"

"So if we add up all the numbers from one to nine and then divide by three for the three rows, we'll know the sum of each row. "

"Great idea!" said Elizabeth. "So one plus two is three, add three more to get six, add four to get ten…"

"Forty-five," said Lulu, "is the answer." Seeing the perplexed look on her sister's face, she explained how she was able to do the calculation so quickly. "1+9, 2+8, 3+7, and 4+6 all add up to ten. That makes forty. Add the five in the middle and we have a grand total of 45," said Lulu.

Elizabeth could only manage a faint "Wow…"

"Thanks," beamed Lulu, "but that's nothing. Remember the story about Gauss?"

"Fred Gauss?" asked Elizabeth, thinking of the character in the "Life of Fred" math series (by Dr. Stanley Schmidt).

"Not exactly," said Lulu, "Carl Friedrich ~s, the German mathematician. When Gauss ~ge, his teacher asked him to add up all ~s from 1 to 100. It took Gauss less than a minute to solve the problem. Imagine his teacher's astonishment when he got the correct answer!"

Play Along: (5 pts.) Can you figure out the answer to Gauss's problem (the sum of all the integers from 1 to 100) using the technique that Lulu just described? Hint: What is 1 + 100? 2 + 99? How many such pairs exist? Keep reading for the answer.

"You mean 101 times 50, or 5050? Interesting, but let's get back to the problem," continued Elizabeth, "Forty-five divided by three is fifteen, so each row, column, and diagonal will add up to fifteen. It's easy from there."

"Fifteen sum on a turtle's back.   Yo-ho-ho, and a big bowl of yum!" sang Lulu

"That song is actually appropriate," said Elizabeth. Lulu raised an eyebrow. "No, not the original pirate song from *Treasure Island*, of course," Elizabeth explained, blushing. "I only meant that this is a sort of treasure hunt." Lulu only smiled back at her. "But a turtle's back is actually called a carapace," Elizabeth added quickly.

Lulu sang again, this time at the top of her voice, much to Elizabeth's embarrassment.

"Fifteen sum on a turtle's carapace-
Yo-ho-ho, and a big bowl of yum!
Will send us off to a magical place.
-ho-ho, and a big bowl of yum!"

The girls took another look at the magic square:

The center column already revealed two of the three numbers that must add up to fifteen. The girls began by figuring out the missing number and working from there.

Play Along: (5 pts.) Help Lulu and Elizabeth solve the magic square (copy it onto your own paper). Keep reading for the answer.

8 ⚐	1 ♥	6 ⬟
3 ✳	5 ☾	7 ◆
4 ●	9 ★	2 ✗

As soon as they had finished writing the final number, the entire square glowed.

"The clue said that this magic square could transport us anywhere," recalled Lulu. "I wonder how." She half expected the turtle to crawl off the page and offer them a ride.

"Do you remember when we studied ancient China?" asked Elizabeth.

"Yes, why?"

"...w I remember where I've seen a magic ...e," said Elizabeth, "It's part of a

Chinese legend that is over 3,000 years old. An emperor was walking next to the Lo River wondering how to control the flooding that was occurring. Suddenly he saw a turtle emerge from the river with a magic square carved into its shell. The Chinese people recognize 15 as a special number because there are fifteen days in each of the 24 solar terms in the year. They began using the magic square to control the flooding."

Elizabeth started thinking about how one might use a magic square to manage a flood. Then her thoughts shifted to the Chinese solar year and what happens to the 5.25 days that were missing (15 x 24 = 360, but it takes 365.25 days for Earth to orbit the Sun). She resolved to investigate further when she got home.

Meanwhile, Lulu had not allowed herself to get sidetracked. She squinted at the magic square, looking for a pattern that may reveal the next clue. The square's glow dimmed gradually until only a faint sparkle remained on the numbers the girls had written with the silver glitter pen.

Lulu removed her glasses and rubbed her eyes as she wondered whether the afternoon sunshine was playing tricks on her vision. When she put the glasses back on, she saw that a series of letters and symbols were slowly coming into view beneath the magic square: a new clue!

Play Along

1. (10 pts.) Make your own set of Tangrams. You can use any kind of sturdy paper such as cardstock or a cereal box. With a ruler, draw a small right isosceles triangle (with legs 2 cm long). The corner of the ruler can make the right angle. Both of the triangle's legs should be the same size. Cut out your small triangle, and use it as a template for the rest of the shapes. The final set should include:

    a. Two small right isosceles triangles

    b. One medium right isosceles triangle that can be formed from the two small triangles.

c. One large right isosceles triangle that can be formed from the medium triangle and the two small triangles.

d. One square that can be formed from the two small triangles.

e. One parallelogram that can be formed from the two small triangles.

Save your Tangram set in an envelope or bag. Each chapter in this book has a corresponding Tangram puzzle to solve.

2. (2 pts.) Use your Tangrams to make the image of this bird:

Could it be Mrs. Magpie?

Fun fact: a group of Magpies is called a "parliament" (which is also the term for a group of humans in the British government who gather together to make laws.)

3. (4 pts.) If you had a magic square with four rows and four columns, what would be the sum of the numbers in each row, column, and diagonal? You may use a calculator, but not the "+" button. Hint: What do you notice about the sums 1+16, 2+15, 3+14, and so on. Try using Gauss's method.

4. (6 pts.) Solve this 4 by 4 magic square:

8	11	14	
13			12
	16	9	
			15

Place the numbers from 1 to 16 in the cells, so that every row, column, and diagonal add up to the same number (which you found in the last question).

5. (1 pt.) In the 4 by 4 magic square, which is composed of 16 cells, you can find smaller squares made up of 4 cells like this one:

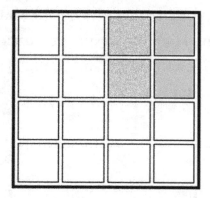

How many such squares (made up of 4 cells) exist?

6. (2 pts.) The 4 by 4 magic square you completed has some extra magic. Pick one of the smaller squares (made up of 4 cells) and add up the numbers. Did you discover what makes this magic square extra magical?

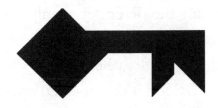

3. Secret Codes

Lulu and Elizabeth stared at the new clue:

Now it's time to get your

$\underline{}\,\underline{}\,\underline{}\,\underline{}\,\underline{}\,\underline{}\,\underline{}$.
✗ ♥ ♦ ♦ ♥ ♦ ☽

Don't forget to pack a

$\underline{}\,\underline{}\,\underline{}\,\underline{}\,\underline{}\,\underline{}\,\underline{}$.
✴ ♥ ✗ ✗ ♥ ♦ ☽

Lulu's eyes lit up. "The symbols are the ones from the magic square," she said, "so this should be easy."

Grasping the silver pen firmly, she wrote the number 2 above the "X" symbol. Before the ink had a chance to dry, the number she wrote began to fade. Soon it disappeared entirely.

"That can't be right," Elizabeth jumped in, "It's obviously a word, and words are made up of letters, not numbers."

"Well, I was close," protested Lulu, "I bet the numbers stand for letters."

"We're not linguistic experts, and there's no key provided," reasoned Elizabeth, "so I'm guessing that this should be straightforward." She grabbed a separate piece of paper and began writing a key.

1 2 3 4 5 6 7 8 9
A B C D E F G H I

"Sometimes the correct solution is also the simplest," Elizabeth said. "Let's start with this and see where we get. I'll do the first word, and you'll do the second one."

Play Along: (4 pts.) Help Lulu and Elizabeth solve the puzzle (on your own paper). Keep reading for the answer.

Both girls worked silently on their tasks and soon they had their answers. Elizabeth wrote her word in the book first. This time, the ink remained bold.

Next, it was Lulu's turn. Keeping her handwriting neat took some effort because she was giggling so hard. "What's so funny?" Elizabeth inquired. When she saw the word Lulu had written, she too burst into laughter.

"Does that say what I think it says?" Elizabeth asked, narrowing her eyes and tilting her head in disbelief.

> *Now it's time to get your*
> BAGGAGE.
> ✖ ♥ ♦ ♦ ♥ ♦ ☽
>
> *Don't forget to pack a*
> CABBAGE.
> ✴ ♥ ✖ ✖ ♥ ♦ ☽

"Yep. Now it's time to get your baggage. Don't forget to pack a cabbage," responded Lulu between giggles. She recited:

> *"'The time has come,' the Walrus said,*
> *To talk of many things:*
> *Of shoes - and ships - and sealing-wax -*
> *Of cabbages - and kings -*
> *And why the sea is boiling hot -*
> *And whether Pigs have wings.'"* [1]

"It's Lewis Carroll," Lulu said. "The only poem I know with a cabbage in it. The Walrus and the Carpenter trick these poor little oysters into becoming their dinner. The moral is to be careful whom you trust. I wonder how we'll use the cabbage on our adventure."

Elizabeth appeared to be ignoring her. "If a word has seven letters," she said, "and each of those letters can be represented by any of the nine symbols on the magic square..." Elizabeth grabbed a calculator out of the backpack.

"So there are 9x9x9x9x9x9x9 or 9 to the power of 7 different possible words. Do you know how many words that is?" Elizabeth frantically entered the exponent on the calculator. "It's 4,782,969! More than four million words to choose from and the answer is 'cabbage'? Really? Unbelievable!"

Lulu didn't understand why her sister was getting so flustered. "I'm sure most of those four million permutations don't form real words. Each syllable needs a vowel, you know," she said as she began walking backward.

"Where do you think you're going at a time like this?" Elizabeth demanded.

"To our garden," Lulu replied. She turned around and began sprinting.

"Why?" Elizabeth yelled after her.

Lulu yelled back, mid-run, without turning around, "To get a cabbage."

This time, Elizabeth did not chase after her sister. *What would the neighbors think?* Instead, she followed Lulu in a rapid walk that soon broke into a type of gallop. She was sure she looked ridiculous, but concerns about neighbors and appearances were soon trumped by curiosity.

When Elizabeth arrived in their family's backyard garden, she found her sister holding a large round cabbage. Elizabeth shook her head at the sight of Lulu's dirt-covered hands. "What now?"

Lulu smiled and raised the cabbage high above her head, sending little chunks of soil flying into her hair. "We have procured the cabbage!" she yelled at the top of her voice.

"Not again," Elizabeth mumbled under her breath.

The distant cry of a crow could be heard in response. A gust of wind suddenly sent aspen leaves showering down around them. Lulu, her cabbage still held high, spun on her heels.

Leaves, dirt, and hair were all flying wildly about. Elizabeth covered her face with both hands.

In a minute, all was calm. The wind had died down to a gentle breeze. Elizabeth removed her hands from her face. A carpet of yellow leaves now covered the ground. As Elizabeth scanned the leaves, taking a mental picture for a future painting, something shiny caught her eye. Moving aside some leaves with the toe of her shoe, she saw what appeared to be two silver tickets. Without haste, Elizabeth picked them up.

Lulu brushed hair and dirt from her eyes and craned her neck to look. "What did you find?" she asked.

"Tickets. They weren't here before. The wind must have blown them in." Elizabeth examined them more closely. "Tulgey Wood Guided Tours," she read aloud. "Where have I heard that name before?"

"Lewis Carroll," Lulu promptly responded.

"What?" Elizabeth looked up.

"Lewis Carroll. He wrote Alice in Wonderland and lots of nonsense poems.

Remember Jabberwocky?" Lulu snatched one of the tickets from Elizabeth, waved it in one hand (as the cabbage was still cradled in the other), and recited:

> *"And, as in uffish thought he stood,*
> *The Jabberwock, with eyes of flame,*
> *Came whiffling through the tulgey wood,*
> *And burbled as it came!"*[1]

"I'm not sure I like the sound of that," said Elizabeth, "I'm glad it's just a poem- completely unrelated to these tickets, I'm sure. I wonder if there is an explanation in *Mrs. Magpie's Manual.*" Elizabeth was surprised at her own words. *Did she really believe that Mrs. Magpie had sent them the tickets?*

"First, grab my backpack," said Lulu.

Elizabeth reluctantly complied, holding the bag open so Lulu could place the cabbage inside. "Are you really going to keep that cabbage in your backpack?" she asked, "It's going to rot and smell awful."

"We'll need it for the adventure," Lulu replied, brushing the dirt off the front of her shirt.

Elizabeth took one look at her sister's fingernails and decided that she would be in charge of the book from now on. She opened *Mrs. Magpie's Manual* and looked down at the next page. It contained a blank square and a rhyme.

Don't just sit there quiet as crickets,
Now is the time to activate your tickets.
In this box, without hesitation,
You must draw the symbol of your destination.
Tulgey Wood is 2C if 2B is the middle,
Don't understand? Then solve this riddle:
With this odd number that you can believe in,
Take one away and you've got an even.

"So it sounds like we have to draw a symbol in the box," Elizabeth said.

"But which one," chimed in Lulu. "Do you understand the riddle? Any odd number becomes even when you add or subtract one from it. Besides, it says to draw a symbol not a number in the box."

Both girls sat silently pondering the clue. The warmth of the sun was pleasant on their shoulders. A cricket chirped an autumn song. Lulu giggled at the irony.

"The magic square can transport you anywhere," she suddenly blurted out.

"What?" asked Elizabeth.

"Remember the clue about the magic square? It said that it could transport you anywhere. Maybe it's the key to all this."

Lulu flipped back to the magic square. "2C and 2B sound like coordinates on a map. Maybe the magic square is a sort of map." She took a pencil out of her backpack and began labeling the rows and columns of the magic square:

	A	B	C
1	8 ▸	1 ♥	6 ⬟
2	3 ✹	5 ☽	7 ◆
3	4 ●	9 ★	2 ✗

"Look!" exclaimed Lulu. "Row 2, Column B, or coordinate 2B, is the middle of the square. I think you're on to something."

Play Along: (1 pt.) The clue said that the symbol for Tulgey Wood is found at coordinate 2C. Which symbol should Lulu and Elizabeth write down? Keep reading for the answer.

"So 2C is the number seven," continued Lulu. She paused for a moment and then laughed out loud.

"What is it?" Elizabeth asked.

"Seven. Take one away and it becomes even," Lulu said, still laughing.

"Seven take away one is six, so what?" Elizabeth felt herself getting frustrated that Lulu was not letting her in on a private joke.

"No, take away one *letter* to make it even," explained Lulu. "Take the letter 's' away from seven and you get the word 'even'. Get it?"

Elizabeth sighed. "Mrs. Magpie has a strange sense of humor," she grunted.

"I think we need to draw a diamond since that is the symbol for 7 in the magic square," said Lulu.

Sensing that her sister was trying hard to mask her vexation, Lulu handed the silver pen to Elizabeth.

"You're the artist," she smiled kindly.

Elizabeth was firmly immune to flattery, but, nevertheless, she took the pen and drew a neat diamond in the square on the paper. It glowed as soon as she had finished.

"Look!" gasped Lulu. She held out her ticket. It, too, was glowing.

Play Along

1. (2 pts.) Using your Tangram set (from Chapter 2), construct the following puzzle of a key:

Fun fact: The Rosetta Stone is the most famous "key" to understanding Ancient Egypt. This tablet, which contained the same text in three different languages, allowed archeologists to read hieroglyphics. The task of deciphering its code, however, was much more difficult than Lulu and Elizabeth's puzzle and was not completed until 23 years after the Rosetta Stone's discovery.

2. (2 pts.) Use the magic square from the chapter to decode this silly word:

3. (4 pts.) A permutation is a big word that just means a combination of objects where the order matters. For example, the word BAD and DAB have the same letters but are different permutations of those letters.

Elizabeth calculated the number of permutations of seven-letter words with nine possibilities for each letter. The technique may have sounded complicated, but it's not that difficult. Let's say you want to calculate the number of three-letter words that can be made using the nine letters from the magic square. The first letter in the word has nine possible values (A, B, C, D, E, F, G, H, and I). So does the second letter and the third letter. So just multiply: 9 x 9 x 9, and you get 729 possibilities (or permutations).

Now how many three letter words (other than acronyms) do you know that don't have any vowels (A, E, I, O, U, and sometimes Y)? None. Lulu made a good point when she said that real words have a vowel sound in each syllable. We are more likely to come up with real 3-letter words if they follow the CVC (consonant-vowel-consonant) pattern. Can you calculate the number of permutations of those types of words that exist using only the first nine letters of the alphabet? (Hint: The middle letter can only be A, E, or I). Feel free to use a calculator.

4. (1 pt.) Lulu said that any even number becomes odd when you add or subtract 1 from it. Is she correct? What happens when you add or subtract 1 from an even number?

5. (2 pt.) Figure out what happens when:
      a. You add two even numbers
      b. You add two odd numbers
      c. You add an even and an odd number

6. (1 pt.) Look at the grid created by the magic square that Lulu and Elizabeth solved:

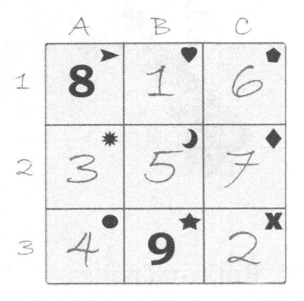

Draw the symbols for each of these cells: 1C, 3B, and 2A.

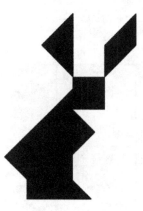

4. Rabbit Trails

The girls clutched the glowing tickets and looked at each other.

"What do we do with these?" asked Elizabeth.

"I guess we should turn the page and solve the next puzzle," Lulu answered.

Elizabeth let out a loud sigh.

"Aren't you having fun?" Lulu asked.

"I'm enjoying the puzzles if that's what you mean," responded Elizabeth. "But it feels like we are in a labyrinth – a giant maze with twists and turns that never actually leads anywhere."

"I think we're close. I can feel it. How do you explain the glowing tickets?"

"Decoys," said Elizabeth skeptically. "Maybe this whole thing is just a series of tricks to entice us to continue. Glowing does not require magic. I could probably concoct a chemiluminescent mixture with my chemistry set. Or buy a glow-stick at the dollar store."

"Mrs. Magpie is an honest lady, or bird, or whatever she is. She has no reason to deceive us," retorted Lulu.

Elizabeth opened her mouth, but before she could get a word out, Lulu continued. "I know what you're going to say, and you're right. I don't know Mrs. Magpie and there is no empirical evidence that this book is magical or will lead us to another world.

Elizabeth listened, silent and still.

"But I have this feeling," Lulu continued, "in my heart that Mrs. Magpie is kind and this book is indeed magical, and that our grand adventure is just around the corner. It's not scientific, I know, but sometimes you must follow your instincts."

Lulu was relieved by the silence that ensued. She was not in the mood for an argument. Her head spun with ideas for what to say next. But as soon as she turned the page, her face lit up.

"I don't care if you agree with me," Lulu said, "but the next challenge is made for you. Look at this."

Both sisters bent over the page to look at the next clue.

A task impossible for a mortal,
Draw a perfect circle portal.

Elizabeth closed her mouth and smiled. "I can draw a perfect circle," she bragged.

Snatching the silver glitter pen, Elizabeth began drawing a circle. As soon as she took the pen off the paper, her circle faded and disappeared. Tenacious as always, Elizabeth immediately tried again.

Lulu didn't say anything. Her sister's circle was splendid. In fact, it was more precise that anything that she could draw herself. But it wasn't perfect.

"Nobody can draw a perfect circle," Lulu murmured.

"Giotto could," said Elizabeth. She was working carefully on her fourth attempt at a perfect circle.

"Giotto?" asked Lulu.

Elizabeth looked up from her work. "He was an Italian painter. His submission to the Pope to work on St. Peter's Basilica was a single circle drawn freehand in red paint. The Pope was so impressed that he gave Giotto the job."

"You're a great artist, too," said Lulu in her kindest voice. "I just don't think Mrs. Magpie intends for us to draw the circle without using a tool."

Elizabeth did not look convinced. She was now pressing her pinky down in the center of the paper, with the pen between her index finger and thumb. Like a human compass, she rotated the entire book with her other hand. "Steady. Steady," she said aloud. The resulting circle appeared almost perfect, but it, too, disappeared before the ink had dried. Elizabeth sighed.

"The clue said we were drawing a portal. That means a door or gateway. What if it will lead to a different world? Aren't you just a bit curious?" Lulu asked.

Elizabeth bit her lip and put down the pen. "Hand me the compass," she said with a sulk.

Lulu reached into her backpack. The cabbage took up most of the space. She took out some more pencils and felt all around for the compass. It wasn't there.

"The compass isn't here," Lulu told her sister.

"Should we head inside and look for it?" Elizabeth wondered.

The sky was already beginning to darken, and the breeze had a chill to it.

"There isn't any time," Lulu replied. "We'll just have to make our own compass."

Elizabeth picked up a pencil and Lulu tied her hairband to it. The girls fastened the other side of the hairband to the silver pen. Lulu held the pencil steady in the center of the page while Elizabeth spun the pen around it, being careful to keep the hairband outstretched so the radius of the circle would be the same at every point.

No sooner was the circle completed before its entire area changed into a light brown color that resembled the texture of wood grain. Lulu's first impulse was to knock on it. A hollow sound echoed from the book.

The twins looked at each other. "Is it a door?" Lulu whispered.

"A door for a squirrel, maybe. There's no way we'd fit through that tiny circle," said Elizabeth.

"Unless we find a potion that says 'Drink Me' which will make us smaller," Lulu thought aloud.

Elizabeth rolled her eyes again. "You're not Alice in Wonderland," she said, pushing against the little door with the palm of her hand while holding the book firmly in the other.

Lulu smiled to herself, pleased that her sister was showing interest in the door and the adventure that surely lay ahead (even if she refused to admit it). Lulu recited:

"And when I found the door was locked,
I pulled and pushed and kicked and knocked.
And when I found the door was shut,
I tried to turn the handle, but-"[1]

"But what?" asked Elizabeth, taking her hand off the door.

"It's Humpty Dumpty's song. It just ends that way," replied Lulu. "Wait- *I tried to turn the handle'.* Do you suppose....?"

Before Lulu could complete her sentence, she caught a glimpse of the open page of *Mrs. Magpie's Manual* and pointed. Elizabeth understood instantly.

The twins realized that they had been so busy examining the round door and conjecturing about where it might lead, that they hadn't noticed the new clue that had appeared further down on the page.

No one can leave, and no one can enter,
Through a door without a knob in its center.

"All the other clues gave us instructions, but this one is rather vague," Lulu observed.

"Let's try drawing a doorknob," said Elizabeth. She picked up a pen and drew a very convincing doorknob on the circular door. Her markings, which were already faint on the wood, immediately vanished.

"Not again," Elizabeth said with a deep sigh.

"A knob in the *center*," said Lulu. "That's what the clue says. Do you know how to find the exact center of a circle? If I've learned anything from these clues, it's that we must be precise. Mrs. Magpie is very particular about the solutions."

Elizabeth's face glowed. "This, I can do!" she exclaimed. "I can actually find a circle's center in a number of different ways."

"One way will be enough," said Lulu.

Elizabeth reached for the ruler and handed Lulu a pencil. "Here's what we do," she began in a pedantic tone.

Play Along: (8 pts.) Trace a circular object onto your own piece of paper. Grab a ruler, and follow Elizabeth's instructions for finding the center of your circle.

"Pick two points along the circumference of the circle. Any two points will do."

"Circumference? Oh, the outer edge of the circle- the perimeter. Of course." Lulu drew two points along the circle's perimeter.

"Now we connect them," said Elizabeth as she used the ruler to draw a straight line between Lulu's points. "This is called a chord."

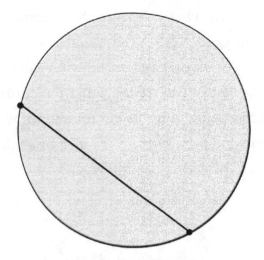

"Next, we need to bisect the line," Elizabeth continued.

"Dissect it?" Lulu asked. She had dissected a frog last summer at a science camp for gifted children. The frog's digestive system was still in the family's freezer, in fact.

"No, *bisect* it. Split the line exactly in half," Elizabeth explained. She measured the chord and found that it was 8 centimeters from end to end. Dividing it in two was easy. She made a small dot at 4 centimeters.

"We can't just draw any line through the center of the chord; it must be perpendicular which means-" Elizabeth started.

"I know what perpendicular means,"
blurted out Lulu. "It's a right angle like the
corner of a book or... the corner of a ruler!" She
placed the end of the ruler against the chord so
its corner was touching its center and drew a line.
She even added the symbol for a right angle – a
little square placed at the intersection of the two
lines.

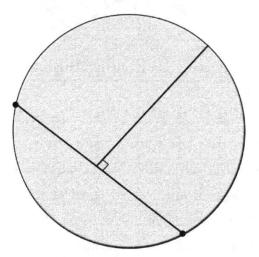

Elizabeth took the ruler and extended the
perpendicular line until it reached from one end
of the circle to the other. "This is a special type
of chord," she said, "because it passes through
the center of the circle. It's called the circle's
diameter."

"But the two points I chose for the first chord were random. Will this work with any two points?" asked Lulu, amazed.

"Yep. It sure will. Let's measure the diameter and find its center," continued Elizabeth.

"Which will also be the circle's center," said Lulu with excitement. She was so caught up in geometry that she had forgotten for a moment that this circle was a door and they were getting close to figuring out the clue.

Elizabeth determined that half of the 14-centimeter diameter of the circle was 7 centimeters. "The radius is 7," Elizabeth mumbled to herself as she placed a dot in the center of the diameter using the greatest care.

As Elizabeth marked the center of the circle, both girls held their breath. They did not have to wait long before the dot transformed into a small doorknob.

Lulu and Elizabeth stared at the doorknob that had just appeared on the small wooden door.

"Should we open it?" Lulu asked softly.

She suddenly noticed that a new clue had replaced the old one.

Galileo, Newton, Descartes, and Pascal too,
All followed rabbit trails that led to something new.

"Rabbit trails?" asked Lulu.

"You know, like when one idea leads to another," said Elizabeth. "Mother lets us follow rabbit trails all the time for homeschool. Whenever we have a question, we dig deeper and see where it takes us."

"Oh yeah. But there's no more time to waste," responded Lulu with resolve. The sun now hung low on the horizon. She grasped the tiny doorknob in her fingers and pulled.

The miniature door, which indeed had begun as a mere circle drawn in a book, swung open. Lulu held the book close to her face so she could peer into the opening. All she could see was some sort of a tunnel or hole.

"Let me see," Elizabeth demanded, snatching the book. She, too, peered through the open door but could only see a dark corridor. She shook the book to see if anything would come out, but had no success.

Without consulting her sister, Lulu reached inside the tunnel in the book that Elizabeth was holding. Her arm was already in up to her elbow when she let out a yelp. Lulu quickly pulled out her arm and stumbled backward in the grass.

Elizabeth dropped the book in alarm and rushed to her sister's side. "What happened?" she asked Lulu, taking her arm.

"At first, I thought the tunnel was empty," began Lulu, regaining her composure, "but then I felt something soft and warm. I wanted to pull it out of the tunnel so we could see what it was, but as soon as I grabbed it, something scratched me. It didn't hurt, but sure gave me a fright." She stretched out her arm to Elizabeth revealing a small red scratch.

"That's enough," said Elizabeth, trying to hide the distress in her voice. "It's getting late and who knows what is behind that door. What if it's dangerous? We just can't risk it." She reached for the book which had fallen face down in the grass with every intention to shut it forever.

Before Elizabeth was able to grab the book, however, its pages began to rustle. "Look!" she yelled out. Both girls watched as a small nose poked out from beneath the book, followed by the furry brown body of a rabbit. The rabbit's whiskers twitched for a moment before he bounded away through the garden and beneath the gate.

"Follow the rabbit!" shouted Lulu. Elizabeth opened her mouth to protest, but Lulu had already taken off after the little critter.

Play Along

1. (2 pts.) Using your Tangram set (from Chapter 2), construct the following puzzle of a rabbit:

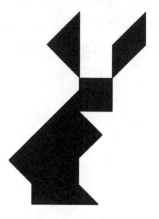

Fun fact: A male rabbit is called a buck and a female rabbit is called a doe, just like deer. Baby rabbits, however, are not called fawns (or bunnies, as many people think); a baby rabbit is called a kit, short for a kitten.

2. (1 pt.) How close can you come to drawing a perfect circle without using any tools? Find a circular object that can be traced, like a jar lid or

cup, and place it on your paper. Mark a couple points around the perimeter of the circle. Then remove the object and try to draw the circle that those points lie on, without using any tools. When you've finished, trace your object using a different colored pen or pencil to see how close you came.

3. (2 pts.) Create your own compass by connecting two pencils with a string. Hold one of the pencils still while you rotate the other one around it. If you keep the string stretched, you'll draw a perfect circle.

4. (5 pts.) In the chapter, did you trace a circle and follow Elizabeth's instructions to find its center? Now measure the circle's diameter, as well as its circumference. Use a calculator to divide the circumference by the diameter. In math, there is a special number called Pi, which is approximately 3.14 (You memorized more digits than this in the first chapter). How does the ratio you calculated compare to Pi?

Just for fun, here's a Pi rhyme (author unknown) to help you remember the relationship:

> *If inside a circle a line,*
> *Hits the center and goes spine to spine,*
> *And the line's length is "d",*
> *The circumference will be d times 3.14159*

5. (10 pts.) One of Mrs. Magpie's clues mentioned Galileo, Newton, Descartes, and Pascal. Find out more about these famous men. Can you match each one to his "rabbit trail?"

   A. Watching a fly on the ceiling and trying to describe its position led this mathematician to develop the Cartesian plane.

   B. Observing swinging lamps in the Cathedral of Pisa inspired this man's discoveries on the properties of pendulums.

   C. This man decided to give up mathematics. When he had a bad toothache, however, he instinctively thought about a math problem to take his mind off the pain. The toothache immediately vanished. He took this as a sign to continue pursuing the problem.

D. Seeing an apple fall from a tree (not necessarily on his head, as legend tells), this scientist began wondering why an object always falls in a path perpendicular to the ground. The apple led him down a rabbit trail which helped him to understand the force of gravity.

5. Two Worlds Join

Elizabeth shook her head, picked up the book, slung the backpack on her shoulders, and followed her sister back across the street to the park. When she had caught up to Lulu, Elizabeth found her twin sister reciting Pi beneath the same maple tree where *Mrs. Magpie's Manual* had been found. *The tree with the Kaibab squirrel,* she recalled. The rabbit was nowhere in sight.

"Where did he go?" Elizabeth asked once she had caught her breath enough to speak.

"I know you won't believe this, but he jumped up and hit the first branch of the tree. Then a door opened up in the trunk, and he went in. It all happened so quickly," Lulu rambled excitedly.

"Do I need to remind you again that you are not Alice in Wonderland?"

"I'm telling the truth," Lulu insisted, "The clue told us to follow rabbit trails to discover new things."

"You take everything so literally," said Elizabeth. "Besides, this is no white rabbit you're following. Did you notice the smaller ears? I think it might be a cousin of the cottontail that can swim. It's called a marsh hare."

"The March Hare?"

"No, a *marsh* hare or, more accurately, marsh rabbit," lectured Elizabeth, looking up at the darkening sky. "The strange thing is that they live in Florida, nowhere near here, and-" When Elizabeth looked down, Lulu was already hanging from the first branch. Much to Elizabeth's relief, nothing was happening.

"I'm sure this is how he did it," Lulu said to herself.

Lulu released her grip on the branch and, down on hands and knees, began feeling around the base of the tree for an opening.

"What's this?" asked Elizabeth pointing to a symbol that someone had carved into the trunk. "I don't remember it from when we were just here."

Lulu stood up and ran her fingers along the carved symbol. "Is it a sideways eight or the sign for infinity?" she wondered. "I bet the book will tell us!"

Elizabeth opened up *Mrs. Magpie's Manual* and saw that a new clue had appeared beneath the one about rabbit trails. She read it aloud:

Before you enter Wonderland,
Take the ticket in your hand,
And as you wrap it 'round your wrist,
Give the strip a little twist.
Once this simple deed is done,
Our two worlds will join as one.

Elizabeth could see Lulu's smile broaden at the mention of Wonderland. "I guess you *are* Alice, after all," she conceded.

"I would let you be the Cheshire Cat, but you don't smile nearly enough," Lulu replied.

"I don't need to be anyone," blurted out Elizabeth, surprising herself. She suddenly realized how much she truly meant that statement. "Let's do this."

Lulu glanced at the clue again. "The symbol isn't an infinity sign, it's a Mobius strip," she said. "I think I understand!" She held up her silver ticket, which reflected the golden orange sunset. "If I put the two ends of the ticket together like a bracelet-" Lulu stopped mid-sentence. The ticket seemed to be magnetic; its two sides stuck together to form a loop.

After a moment, Lulu continued. "Let's pretend that Earth is on the outside of the loop, and Wonderland is on the inside. The two worlds will never meet." Lulu traced her finger along the outside of the loop and returned to where she had started without ever touching the inside of the loop.

"I see!" Elizabeth chimed in. "If we put a twist in the loop like this-" She took her own ticket and formed a loop with a single twist in it.

"-then you have a Mobius strip," Lulu completed her sister's sentence. She created one with her own ticket and once again began tracing her finger along the outside of the ticket. Without lifting her finger, she was soon running it along the inside surface of the ticket. "Two worlds join as one."

The twins put the Mobius-strip tickets around their wrists. "Ready?" Lulu asked as she reached for the first tree bough. Elizabeth nodded. As Lulu pushed down on the branch, the girls heard a faint cracking noise. Upon further examination of the tree trunk, they saw a narrow slit that had not been there before. Lulu slid her fingers into the crack, which swung open with a sigh. In another instant, she had disappeared into the tree.

Elizabeth took a deep breath and followed her sister. She clenched her fists expecting to fall (*wasn't that the way to Wonderland?*), but instead found herself standing in darkness.

"Lulu? Lulu?" Elizabeth's voice sounded hollow. No reply came. Elizabeth frantically felt around in the darkness. She could no longer find the door. In fear and desperation, Elizabeth began lunging with her entire body against the smooth walls of wherever she was. All of a sudden, the wall gave way and she found herself lying down on green grass and blinking in bright sunshine.

Elizabeth instantly remembered that it was early evening when she had entered the tree. *Why was the sun suddenly so bright? Where was she?*

"Elizabeth!" She recognized Lulu's voice. Soon the figure of her sister was standing over her, and a familiar slender hand was helping her get up.

"Welcome to Tulgey Wood!" exclaimed Lulu with a broad smile. "I got here just in time to see the March Hare swim across that river. She pointed to a river that was flowing just a couple paces away from where they stood.

"*Marsh* hare," corrected Elizabeth. She looked around. At first, as she scanned the landscape, Elizabeth did not spot anything that would suggest that they were in a different dimension. Tulgey Wood was far from the dark, dense forest that Elizabeth had imagined. Yellow, orange, and red leaves decorated tall trees. Gaps in the colorful canopy revealed a bright blue sky. Still, she felt deep in her stomach that something was just not right.

"Isn't it picture-perfect?" asked Lulu with a contented sigh. She began skipping around the trees, stopping here and there to pick up leaves.

That's just it, thought Elizabeth. *This place is almost too perfect.* She instantly realized that the surreal perfection of the landscape was the source of her unease.

Elizabeth studied the tree next to which they were standing. A large trunk divided into two identical large branches like the letter "Y". Those split into two more branches. The process went on and on, symmetrically, up through to the thinnest twigs which held the colorful leaves.

It's a fractal tree, thought Elizabeth. *The pattern repeats so if I break off any branch of the tree, it will look like a smaller version of the entire tree.*

She looked down at the ground and picked up a handful of fallen leaves. They, too, had something unusual about them.

Elizabeth took the stems off her leaves and arranged them on the ground.

The leaves form a tessellation, she thought. *Are we in an M.C. Escher drawing? What if we're stuck here forever?* Panic ran through Elizabeth's body.

Lulu strolled over to her sister with a crown of leaves around her head. She glanced at the leaf tessellation on the ground. "That's pretty," Lulu remarked and continued the pattern until a yellow and orange rug spread before them, each leaf fitting perfectly into the next.

Her sister's presence gave Elizabeth some comfort, but she was still worried. "How do we get back home?" she asked aloud. She couldn't quite share her sister's enthusiasm for this place.

"We probably have to complete a perilous quest, first," guessed Lulu.

How could my sister be so nonchalant about the situation? Elizabeth wondered. "We should consult Mrs. Magpie's book for further instructions," she said. Elizabeth was glad that she had the foresight to bring the book. *What would her sister do without her?* She opened it, quietly praying that their mission would not involve any dragon-slaying or giant spiders.

A Jubjub bird disguised as a lark,
Borogroves concealing a Snark,
When you're in Tulgey Wood, you must
Be careful whom it is you trust.
But regardless of where you decide to roam,
You're only one tree away from home.

"Jubjub bird? Snark? I guess this place is not as innocuous as it seems," whispered Elizabeth.

"What does *'one tree away from home'* mean?" Lulu wondered. She walked over to a tree on the riverbank and pushed on its lower branch. A crack appeared in the trunk – another door.

"That takes all the fun out of the adventure," Lulu said with disappointment. "If every tree in this place is a way home, then we're never in any real danger. That sucks all the excitement out of it, don't you think?"

Elizabeth was secretly relieved. "On the contrary," she said, "it creates an even bigger challenge.

Lulu looked puzzled. "What do you mean?"

"Well," continued her sister, "it's harder to stick with something if the temptation of an 'easy out' is lingering over you. It's like solving a complicated math problem. You know you can quit at any time (or read the answer in the back of the book), but you keep trudging through to get to that gratifying state of Q.E.D."

"Q.E.D.?"

"Quod erat demonstrandum. It means that you demonstrated what you had set off to show. Mathematicians put it at the end of their proofs. It's the 'ta da!' of math. I thought you'd know that since you've been learning Latin."

"Fair enough," said Lulu thoughtfully. "Let's get this adventure started."

<u>Play Along</u>

1. (2 pts.) Using your Tangram set (from Chapter 2), construct the following puzzle of a Mobius strip:

Fun fact: The Mobius strip was discovered (and named after) August Ferdinand Mobius, a German mathematician and astronomer.

2. (3 pts.) Take a piece of paper and cut it lengthwise to create two long strips of paper, approximately an inch thick.

     a. Tape the ends of the first strip together to form a loop.

Using a marker, draw a line that starts on the outside of your loop. Did your marker ever touch the inside of the loop?

b. Give the second piece of a paper a half twist before taping the ends together to form a Mobius strip.

Start making a line with your marker on the outside of the Mobius strip. What happens?

3. (3 pts.) Cut both the loop and Mobius strip that you made in half (along the line made by your marker). What happens? Use your results to fill in the missing words in this limerick (from an unknown source):

4. (2 pts.) Cut out another strip of paper and draw two lengthwise lines on it (dividing the paper into three long sections). Then turn this strip of paper into a Mobius ring by giving it a half-twist and taping the ends. Cut the strip along the lines you drew. What happens?

5. (5 pts.) Fractals are figures with patterns that continue repeating so that even the smallest part resembles the whole. You can find fractals in nature; some examples are snowflakes, crystals, and pineapples. However, trees in nature do not exactly follow the pattern that Elizabeth described. Draw a picture of the fractal tree from Elizabeth's description. Then, look at some

trees in nature to see how they differ from your drawing.

6. (8 pts.) A tessellating shape can be repeated over and over to cover an area without any gaps or overlapping (a tiled floor is an example).

Design an original tessellation. Find a piece of thick paper or cardboard (cardstock works well), and cut it into a small rectangle. Cut out a shape from one side of the rectangle. Tape the shape that you cut out to the opposite side of the rectangle. Repeat until you are happy with the result.

Step 1   **Step 2**   **Step 3**

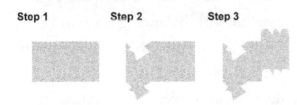

On a separate piece of paper, trace your tessellation over and over to cover your paper. Color in your work. Look at some M. C. Escher drawings for inspiration.

6. River Crossing

"Tickets, please," said a voice, startling the twins. They turned around and found, to their surprise, that the owner of the voice was a Pig. He was standing upright, balancing his plump body precariously on two hooves. The Pig wore a shirt with a frilled collar and an oversized purple bow-tie. "Tickets, please," he repeated.

After the shock of seeing a talking Pig had worn off, the girls remembered the silver tickets around their wrists, which they extended toward the Pig. He inspected the tickets, snuffling with his snout a little too close for comfort, in Elizabeth's opinion. She happily snapped her hand back to her side when he appeared to have finished.

"The name is Pepper Pig," the Pig announced cheerfully, "Tulgey Wood tour guide extraordinaire!" He gave a little wobbly bow and sneezed.

"I was born and raised here. Nobody knows these woods better than I."

"He even has proper grammar," Lulu whispered to Elizabeth, "Most people would say 'better than me.'"

Pepper Pig, who had heard the compliment but pretended that he hadn't, smiled and sneezed again. "Mrs. Magpie, the regent of these parts, has requested that I deliver you to her safely. That is just what I'll do," he continued, "as soon as we've settled the small matter of my payment."

The girls looked at each other. "We haven't brought any money," said Lulu meekly.

"Money? What use is money to a Pig?" replied Pepper. "I accept payment in the form of cabbages."

Lulu let out a giggle. Elizabeth gave her a stern look. She didn't want Pepper to feel insulted. Reaching into the backpack, Elizabeth pulled out the cabbage and raised it in the air ceremoniously. Pepper Pig licked his chops.

"Wait a minute," Elizabeth said, drawing back, "we'll give you the cabbage once you've guided us safely through Tulgey Wood. You look like a decent fellow, alright, but we're strangers here, and it's not so easy to tell whom we can trust."

"That sounds fair," agreed Pepper with an earnest look of disappointment.

"Well then," he adjusted his bow-tie with his teeth, "our first order of business is to cross the river."

All three pairs of eyes turned to the water. The river was murky, and the bottom was not visible. Elizabeth shuddered. "How exactly do we cross it?" she asked, wanting to get the affair over with.

Pepper Pig looked around nervously.

"You've crossed this river, before, haven't you?" Elizabeth prodded.

Pepper gulped.

"Well you're our tour guide, right?" Elizabeth couldn't mask the frustration in her voice.

The Pig did not reply. Lulu was glad that Elizabeth had been smart enough not to give him the cabbage right away. She was sure that he would have been off with it by now.

"The book-" remembered Lulu. "Let's look at the book!"

Elizabeth opened *Mrs. Magpie's Manual of Magic*, which she had been clutching this entire time. A new clue had appeared.

In three hours, three men can build three boats.
How long for one man to build one boat that floats?

Play Along: (2 pts.) Help Lulu and Elizabeth solve the clue. Keep reading for the answer.

"One hour?" asked Pepper Pig.

Lulu and Elizabeth had learned that they always had to take their time to think about the problem.

"We have to make the assumption that all three men work at the same rate," said Lulu.

"So each man builds his boat in three hours," Elizabeth completed her thought.

Lulu took out the silver glitter pen and drew an elegant "3" underneath the clue. She quickly remembered her units and added "hours" to the page. Some new words appeared below her answer.

Though it may be small and have a minor leak,
You will find a boat that provides the help you seek.
Only two will fit, and you must keep in mind
Which of the others will be left behind.

"What boat?" Lulu asked. They all glanced up from the book to see a small wooden boat tied to a tree next to the shore. It was bobbing gently in the water. Lulu began running toward it. All of a sudden she halted. The boat was not empty. The silhouette of a furry someone was sitting in it.

He sang in a smooth deep voice:

> *"Ever drifting down the stream-*
> *Lingering in the golden gleam-*
> *Life, what is it but a dream?"*[1]

Lewis Carroll, again, thought Lulu. *We really are in Wonderland!*

"Curiouser and curiouser,[1]" she said aloud, feeling rather witty (and forgetting her grammar lessons).

The furry figure turned around. It was a Wolf wearing a derby hat. "Pleased to make your acquaintance," the Wolf spoke, removing his hat. "I'm Manxome McLay, but you can call me Manny for short.

"Are you the owner of this boat?" asked Lulu boldly. "We need to get across the river."

The Wolf hesitated. "Yes I am," he finally responded. There was another long pause. "Unfortunately, my lack of opposable thumbs renders me incapable of rowing this boat myself. I usually rely on the service of my Gorilla, but he hasn't shown up for work today. You know how unreliable primates can be."

Lulu stared at the small rickety boat. It bared more resemblance to a wooden washtub than any sea-worthy vessel. It was inconceivable that a gorilla would fit in it. *The Wolf was clearly lying.*

Elizabeth and Pepper had caught up to Lulu and were standing next to her. Pepper Pig and the Wolf were glaring at each other.

"Do you know him?" Elizabeth whispered to Pepper.

"Manxome is my longtime foe," he replied with an emphatic sneeze.

"Never trust a Pig," Manxome McLay hissed back.

Lulu, who had already developed a deep distaste for the Wolf, thought Pepper was the more trustworthy of the two. In fact, the evident hostility between them had made her trust the Pig even more. *I guess the foe of my foe is my friend*, she thought, *just like a negative number multiplied by another negative equals a positive.*

"We'd be happy to help you in exchange for the use of your boat, Mr. McLay," Elizabeth said. She shivered at the thought of swimming across that river.

Manxome McLay stepped out of the boat. Lulu noticed that the bottom of the vessel already contained about a centimeter of water.

"Only two of us can fit in the boat at a time," Elizabeth continued, "so we'll have to take a couple of trips to the other side."

"O.K. I'll take Elizabeth over first," said Lulu.

"Don't leave me alone with that Wolf," begged Pepper.

"Well I can take Manxome over first, and you can stay here with Elizabeth until I come back to fetch you," Lulu told him.

It was Elizabeth who objected this time. "What if the Pig steals the cabbage from the backpack if you aren't here to help me?" she protested. Unlike her sister, Elizabeth had decided that Manxome was more believable than Pepper.

"Can't I just take the cabbage to the other side in the boat?"

"No!" yelled Pepper, "it will get wet in that leaky boat if nobody holds it, and Elizabeth is the only candidate for the job since you are rowing and she's the only one left with opposable thumbs. I won't have it any other way."

Lulu put her face in her hands in frustration. *At this rate we'll never get across*, she thought.

"Can my sister and I have a moment to figure this out?" asked Elizabeth.

Pepper and Manxome both nodded to the girls, then proceeded to glare at each other. Elizabeth and Lulu put their heads together and spoke in hushed tones.

"Let's see if the book can help," suggested Lulu. Elizabeth opened the book, but all she saw was the clue they had already read.

Lulu looked disappointed. "I guess this is our puzzle to solve," she said.

"Let's write down what we know," Elizabeth suggested, taking out a separate piece of paper and a pencil. She began writing

1. Lulu needs to get a Wolf, a Pig, and a cabbage (held by Elizabeth) across the river.
2. She can only take one passenger at a time.
3. The Wolf and Pig cannot be left alone together.
4. The Pig and the cabbage (held by Elizabeth) cannot be left alone together.

"A Wolf, a Pig, and a cabbage!" exclaimed Elizabeth. "Why do I feel like I've seen this river crossing problem before? But I think we're missing a goat, or maybe a bag of beans."

Lulu looked at the notes and grinned. "I know exactly what to do."

Play Along: (8 pts.) On a separate piece of paper, draw and cut out figures of a Wolf, a Pig, and a cabbage. See if you can get them across the river (which you can draw on another piece of paper), without breaking the rules Elizabeth wrote down. Keep reading for the answer.

The girls turned back to the Pig and the Wolf. "Pepper, you're coming with me," Lulu announced. The boat wobbled as Pepper got inside. Its bottom filled with more water, but it stayed afloat as Lulu grabbed the oars and rowed them both to the other side.

Pepper Pig got out and waited on the bank as Lulu rowed back. *Pepper won't run away,* she thought, *because he hasn't yet received his cabbage.*

"Manxome, you're up," Lulu said. The Wolf had an easier time getting into the boat than the Pig. Lulu rowed to the other side. There was a look of dread on Pepper's face as they came towards the shore. "I'm not staying here with- with- with- him!" he sputtered between sneezes.

"Don't worry," laughed Lulu as Manxome came ashore, "get back in the boat, Pepper?"

"Back in the boat?" Pepper said, "I don't understand."

"It will all work out; you'll see," Lulu reassured him. Pepper didn't move. "I promise I'll take you back to this side," Lulu tried to persuade him, "after all, Elizabeth and I would be lost without your expert guidance." The flattery didn't work. Pepper sunk his hooves firmly into the soil.

"And don't forget the large, green, crispy, delicious cabbage." *If flattery didn't work, maybe some bribery would do the trick.*

Lulu finally cajoled Pepper, his mouth watering, to get back into the boat, which rocked wildly upon his entry. This time, they left Manxome waiting on the other side. He wouldn't go anywhere without 'his' boat.

After dropping Pepper off at the shore where they had started, Elizabeth hopped on board, holding the bag with the cabbage. The cold water at the bottom of the boat soaked her shoes and socks and added to her apprehension. She didn't say a word until her feet were once again on solid ground.

"Are you going back to get Pepper?" she asked her sister.

"Of course!" Lulu replied as she set off across the river to fetch the Pig. Her arms were feeling fatigued. *I'm getting a better workout than with a rowing machine*, she thought.

With another trip across the river, all passengers had been safely delivered to the other side. The boat now had a considerable amount of water in it.

"We need to get going," said Pepper, uneasily.

"Goodbye and good luck my dear friends," said the Wolf.

"I have a feeling that's not the last we'll see of that fellow," whispered Lulu as they walked away.

The threesome pressed on. "How far will we have to walk to reach Mrs. Magpie?" Lulu asked.

"Not far. Not far at all," said Pepper. They continued walking in awkward silence. Lulu noticed another peculiarity about Tulgey Wood. The only sound seemed to be the crunching of leaves beneath their feet.

"Don't any birds live in these woods?" she asked.

"Not since the Jubjub claimed this as its territory."

Lulu had a million questions but didn't know where to start. Before she could open her mouth, her sister changed the subject.

"Did you say Mrs. Magpie was the Regent of Tulgey Wood?" Elizabeth inquired.

"Yes. We've had a rash of unfit rulers here, the last of which was an entirely irrational Queen. After the Jabberwocky put a gruesome end to her, the Council decided that someone of sound mind should govern until a proper monarch is crowned. As Mrs. Magpie has no ambitions of taking the throne herself, she was appointed Regent."

Lulu perked up. "Tell me more about the Queen," she said.

"In my humble opinion," replied Pepper, "she was horridly unfit to rule. The Queen had an unpleasant disposition and a twisted view of justice. Many heads would have rolled if it weren't for the compassion of the King who was able to pardon many of the Queen's would-be victims. The Jabberwocky, however, proved far less discriminating about his dinner."

"Elizabeth and I have royal blood in our veins, too, you know," Lulu boasted.

"Not this story again," murmured Elizabeth under her breath.

Lulu continued, oblivious to her sister's eye-rolling. "Our father, who was born in England, is related to Ada Lovelace. She was a famous mathematician in her day and helped develop the first programmable computer. There's even a programming language named after her. But the best part-"

"Here we go," Elizabeth groaned.

"The best part," said Lulu, a little louder, "is that she was a real duchess."

"Duchess? Did you say Duchess?" spattered Pepper, suddenly shaking all over. He was clearly distressed.

"No, sorry, I meant to say she was a countess," Lulu quickly corrected herself to the high relief of the Pig, who undoubtedly had a horrible childhood experience with a duchess.

Before Lulu could resume bragging about her royal connections, or pull any more details out of Pepper, they reached a clearing. It contained a single odd-looking tree in front of a lake with water as smooth as glass. Two small islands stood in the lake's center

Please, no more leaky boats, Elizabeth prayed.

Play Along

1. (2 pts.) Using your Tangram set, construct the following puzzle of a boat:

Fun fact: Rowing has been an event in the Summer Olympic Games since 1900. Many centuries earlier, starting in 1315, rowing races were held in Venice. These ceremonial races, called the Historical Regatta, are still held today.

2. (3 pts.) The boat initially contained 1 cm of water. Lulu, Elizabeth, and Manxome each caused the water level inside the boat to rise 0.5 cm per river crossing trip. Pepper Pig's extra weight caused the water level to rise an additional 1 cm for each trip that he made. How much water was at the bottom

of the boat once everyone had crossed the river?

3. (5 pts.) Lulu and Elizabeth faced a similar river crossing problem last summer when they were camping with their mother and father (their two brothers were at Grandma's house). The two children and two adults had to cross a river using a paddle-boat. The boat could only hold a maximum of one adult or two children at once (one child could also paddle alone, but not with an adult in the boat). How did all four of them get across the river?

7. Seven Bridges

Pepper Pig stood still and looked around. "Now there is something critical we need to do here," he said. He let out a dozen sneezes in rapid succession as Lulu and Elizabeth waited patiently.

"I can't seem to recall what it might be," the Pig finally admitted, shifting his weight uneasily between his hooves.

Elizabeth shrugged. "We might as well consult the book. It's the only thing around here that we can rely on." She opened *Mrs. Magpie's Manual.* A map filled the page.

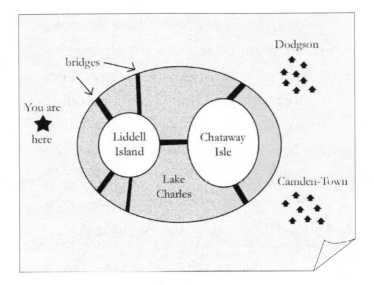

A clue was written under the map:

If you prove yourselves savvy and wise,
A handsome reward in the next town lies.
It's around the lake and over the ridge,
But first you must walk across every bridge.
Fill your bag with plenty of fruit,
As each Bandersnatch will take half your loot.

"Now I remember!" yelled out Pepper excitedly. He gestured towards the strange tree. "This, here, is a Tumtum tree- the one and only."

All eyes turned to the tree. Lulu and Elizabeth now noticed for the first time that its limbs were laden with clusters of small orange fruit.

"The Tumtum fruit is sweet and savory, tart and succulent. It melts the moment it hits your tongue," continued Pepper, his mouth watering. "This makes it a favorite snack of the Bandersnatch."

"Bandersnatch?" asked Elizabeth timidly.

"Yes. A Bandersnatch is a vile creature, I've heard, although I can't say I've ever seen one myself. A very reliable source told me once that it resembles a monstrous crow -as big as a tar barrel, in fact! When I was young, the Bandersnatch gang would take to terrorizing the town as soon as the sun's last rays had vanished. Children who were not back in their homes by dusk would disappear. These circumstances all changed, of course, when Mrs. Magpie made an arrangement with the Bandersnatch leader."

"What sort of arrangement?" Lulu asked.

"The Bandersnatch gang agreed to stop their destructive nocturnal behavior, at least in within town limits, in exchange for a tribute of Tumtum fruit."

"Like the English Kings would pay off the Vikings so they wouldn't attack them," interrupted Elizabeth. "The payments were called Danegeld."

"Umm- yes, just like that," said Pepper, who had never heard of Vikings. "One Bandersnatch lives beneath each of the seven bridges on your map. Each time a visitor passes through Tulgey Wood, he or she must cross every one of those bridges. At each bridge, the visitor must leave exactly half of the Tumtums that he or she is carrying as a tribute to the Bandersnatch."

"We better pick some fruit, then," said Lulu staring up at the heavy boughs of the Tumtum tree. "But how many pieces do we need?"

"So there are seven bridges and at each bridge we lose half our fruit, right?" said Elizabeth.

"Let's make sure we have one Tumtum fruit left at the end," Lulu added, "in case Mrs. Magpie wants it.

"Wait," said Pepper, remembering something important, "no traveler has ever been able to complete the route passing over each bridge only once. They always end up going over one of the bridges twice."

"Then we'll be the first to do it," Lulu dismissed him with a smile.

Play Along: (4 pts.) How many pieces of fruit do Lulu and Elizabeth need to bring in order to have enough fruit for the Bandersnatch under each of the seven bridges, and one left over at the end for Mrs. Magpie? (Keep reading for the answer – cover the next page while you work.)

The twins solved the problem by working in reverse. They started with one fruit and then doubled the number seven times.

"1, 2, 4, 8, 16, 32, 64, 128. We'll need 128 Tumtums," declared Elizabeth when they had completed the calculation.

Lulu stared again at the map. "There is something terribly familiar about that map," she said pensively. "Something is not quite right."

"What do you mean? Our calculation is correct – two to the power of seven is 128," said Elizabeth.

"Which town did you say we were headed to, Pepper? Camden-Town or Dodgson?" Lulu asked.

The Pig went into a sneezing fit. "I... I... I don't know," he sputtered.

Elizabeth rolled her eyes. "I'm sure it will be in the next clue, anyhow," she said.

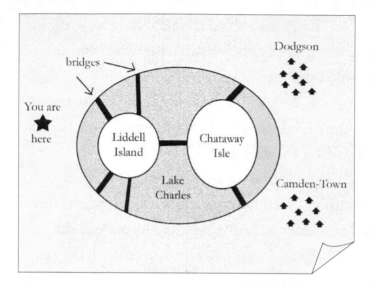

Play Along: (2 pts.) Using your finger, can you figure out a way to cross each of the seven bridges only once? (Keep reading for the answer – cover the next page while you work.)

Lulu tried to trace the route they would take to get to Camden-Town crossing each of the seven bridges once. She tried over and over but could not seem to figure out a way to do it without crossing one of the bridges twice. She then tried finding a route to Dodgson, but could not find a way to reach that town, either without crossing one of the bridges twice.

"Impossible," she mumbled.

"What?"

"Impossible!" repeated Lulu in a louder more confident tone.

Is this coming from the girl who always tells me that nothing is impossible? Elizabeth wondered.

"Now I know why this map looks familiar," continued Lulu. "It's identical to the Seven Bridges of Königsberg problem. Euler was asked to find a parade route through the town that crossed each bridge only once. He proved that it was impossible."

"Let me see," said Elizabeth, examining the map. After a minute of thought, she came to the same conclusion as her sister.

"I guess we'll be making eight bridge crossings."

"That means that we need twice the fruit we'd need if there were only seven bridge crossings. 128 times two is-" started Lulu.

"256!" Elizabeth piped in.

"Now that's a lot of fruit!" exclaimed Pepper.

"The power of exponents!" replied Lulu with a smile. "We'd better get started."

The girls went right to work figuring out how they would pick the Tumtums. Only Pepper stared vacantly at the girls' backpack, drooling. Lulu had pronounced Euler correctly, as "Oiler", which had reminded Pepper how tasty that cabbage would be with a little oil and vinegar drizzled on top.

Lulu and Elizabeth looked up into the tree, then at each other. "The book," they said in unison.

Elizabeth opened the book and wrote "256 Tumtums" under the last clue. New words instantly appeared beneath the number:

To reach the fruit, I'm sure you'll be gladder
To have the assistance of a ladder.
In your world, this notion might seem absurd,
But in this one, each ladder rung is made of a word.
One letter changes, three stay the same,
Now it's your turn to make MATH a GAME.

Lulu looked around for a ladder, but none had appeared. "Math is already just a big game," she said. "We play math all the time."

"I think this puzzle is related to words rather than math," Elizabeth said.

"A ladder is what you need," added Pepper, "a word ladder."

"Did you say word ladder?" asked Lulu. She remembered that she had once played a game with their mother where they took turns transforming one word into another by changing one letter at a time.

Elizabeth had been thinking the same thing. She wrote "MATH" in Mrs. Magpie's book, beneath the clue.

"We need to change the word "MATH" into the word "GAME". They both have the letter "A" in the second spot, so only three letters must change."

"Let's work backward and forwards at the same time," suggested Lulu. "The word that can come before GAME could be GATE. So now we find the word between MATH and GATE. GATH isn't a word, so it must be..."

"MATE!" Elizabeth blurted out. Lulu started laughing hysterically.

"What? It's a noun that means a partner or friend," Elizabeth said without even a chuckle.

"G'day mate," said Lulu in her best Australian accent. She used the pen to finish the word ladder, writing in uncharacteristically neat letters in the book:

```
MATH
MATE
GATE
GAME
```

"That was easy!" Lulu exclaimed. When the girls looked up from the book, a ladder was propped against the trunk of the Tumtum tree.

"It's not very tall," Lulu sighed. She tried to hide her disappointment. The ladder was no taller than she.

"You go," said Elizabeth, who was about as fond of heights as swimming.

Lulu ascended the small ladder. She stretched out one arm while clutching onto the ladder with the other. Only one cluster of Tumtums was within reach. She plucked it and handed it over to her sister.

"Eight," said Elizabeth. "There are eight Tumtums in this cluster."

"Every cluster has eight pieces of fruit, of course," corrected Pepper Pig. "You don't have to count."

Elizabeth grabbed her calculator and divided 256 by 8. "We need 32 clusters," she announced, "31 more to go."

Lulu frowned. The ladder was simply too short to be of any use. "But how do we get a taller ladder? Where can we find the materials?" she wondered aloud.

"*Language is worth a thousand pounds a word,*"[1] Pepper quoted.

"That's it! Words are a valuable resource around here. What we need is a new word ladder!" yelled out Elizabeth. "Each word is a rung. We made a ladder out of four words, so we received a ladder with four rungs."

"But how many rungs do we need?" asked Lulu, "We can't figure out how tall the ladder needs to be without knowing the height of the tree. And we can't measure the height of the tree without a ladder. Sounds like a Catch-22 to me."

"A what?" Elizabeth began. Just then, she spotted her shadow. An idea popped into her brain. "Thales has the answer," she shouted.

"Thales?" asked Lulu and Pepper at the same time.

"Yes, he was a mathematician who calculated the height of the Great Pyramid by measuring its shadow," explained Elizabeth. "We can use the same technique to figure out the height of the tree. Stand here, Lulu, and I'll measure your shadow first."

Lulu stepped off the ladder and stood still while Elizabeth placed her bag on the ground and retrieved the ruler. The ruler was a foot long (12 inches). Elizabeth placed it against her sister's shoe and then made a mark in the dirt at the ruler's end. She moved the ruler and continued making marks in the dirt until she'd reached the end of Lulu's shadow.

"Your shadow is six feet long," she declared.

"That's good to know," said Lulu, "but what does it have to do with the height of the tree?"

"How tall are you?" Elizabeth asked her sister.

"Exactly four feet tall," Lulu began, "I'm petite, but-"

"So your shadow is one and a half times your height," Elizabeth interrupted. "How convenient!"

Lulu verified the figure in her head. Half of four is two. Add that to four and you get six. When she looked up, Elizabeth was already busy measuring the tree's shadow. She sketched a picture of the problem on a piece of paper.

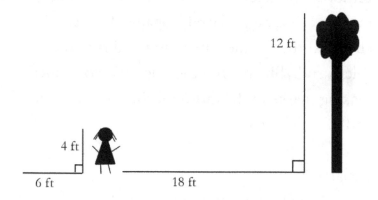

"Don't laugh – it's just a sketch," said Elizabeth, who was very sensitive about her art.

"The shadow of the tree measured 18 feet. When the sun hits objects at the same angle, their shadows will be proportional. I knew that since your shadow was one and a half times your height, the tree's shadow would also be one and a half times as long as the tree is tall.

Lulu nodded her head as she followed her sister's logic. "So the tree must be twelve feet tall since twelve plus six makes eighteen. Thank you, Thales!" she said with a smile.

"Thank you, Thales!" Elizabeth echoed.

Lulu suddenly noticed the Pig's absence. "Where is Pepper?" she asked. Just then, a string of sneezes came from a nearby bush. The girls exchanged glances. They both assumed that Pepper had lost interest and was resting behind the shrub.

Elizabeth measured the distance from the ground to the first rung of the ladder. It was exactly one foot. She usually used metric units, as scientists do, but in this case, she was glad she had measured everything in imperial units (inches, feet, and yards instead of centimeters and meters).

"We need to make MATH into GAME using a total of twelve words," confirmed Elizabeth. She opened Mrs. Magpie's book and opened to the first blank page. "Let's get started."

"MATH, PATH," contributed Lulu.

"Then PITH," said Elizabeth?

"What is pith?"

"It's the spongy stuff inside plant stems."

"OK... Let's see... PITH, WITH," said Lulu as Elizabeth wrote down the words. They continued in this manner until they got back to GATE and GAME. Elizabeth counted the words to make sure there were twelve.

When they looked back at the Tumtum tree, a fine tall ladder with twelve rungs had replaced the shorter one. Lulu began climbing it at once, reasoning that they hadn't any more time to waste.

"Where is the backpack?" she hollered down to her sister, a cluster of Tumtums in her hand.

Elizabeth looked for the backpack where she had left it, under the tree, but it was nowhere to be found. On a hunch, she looked behind the bush where they had heard the sneezes. As Elizabeth neared the bush, she heard repulsive crunching and chewing noises. There was Pepper, his face completely encased in the bag, his curly tail standing upright. *For once, he actually looks like a farmyard pig,* thought Elizabeth.

Elizabeth cleared her throat. The Pig's back end jumped up in the air. After a brief struggle with the backpack, Pepper's face emerged. Pieces of cabbage were stuck to his snout and a pencil hung from his ear. "I was just cleaning out the bag," he explained sheepishly, "to make room for the Tumtums."

Elizabeth gave the Pig a look of contempt and snatched the backpack that was now lying open on the ground. Without a word, she returned to her sister.

"Where did you go?" Lulu asked.

"Never mind. I'll tell you about it later," Elizabeth replied tersely.

Lulu began dropping fruit into the bag which Elizabeth held open. It was no time at all before there were 32 clusters of 8 Tumtums.

"Let's get going," Lulu said. The twins marched ahead down the dirt road. A sullen-looking Pepper trailed a couple of paces behind them.

Elizabeth filled her sister in on Pepper's indiscretions. Lulu thought the sight of the Pig's face stuck in that bag must have been hilarious. She didn't seem as upset by the situation as Elizabeth was. *Will a Bandersnatch eat Tumtums covered in pig germs?* Lulu thought to herself with a chuckle after she realized that nobody had cleaned the bag before filling it with fruit.

Meanwhile, Elizabeth was engaged in more serious thought. "Something has been bothering me," she finally admitted as they walked along. "Do you remember the clue that said, *'If you prove yourselves savvy and wise, a handsome reward in the next town lies.'?*"

"Yes," said Lulu "What do you think it is?" She had already secretly pictured a treasure chest overflowing with gold coins and pearl necklaces.

"I don't know. But I've been trying to figure out why we deserve a reward just for being smart. Aren't rewards supposed to be given out for helping people, creating a useful invention, or something of that nature?" Elizabeth wondered.

"You can use your brain to be helpful. That's the whole point," Lulu replied thoughtfully. "We're not just doing the math to impress others. We're using math as a tool to solve real problems. I know Mrs. Magpie has something more meaningful in store for us."

Lulu and Elizabeth soon reached the first bridge. It was an ordinary, unassuming, wooden bridge. In front of the bridge lay a wicker basket. Lulu pointed to a small sign that was sitting next to it:

> *Kindly leave half your Tumtum stash,*
> *And most grateful will be the*
> *Bandersnatch.*

"He sounds surprisingly polite for such a despised creature," noted Lulu.

"Don't underestimate the Bandersnatch," cautioned Pepper, catching up.

"How does the Bandersnatch know we will be honest and leave half our fruit in the basket?" asked Lulu.

"Not everyone has integrity, you know," added Elizabeth, sending a look of distrust in Pepper's direction.

"Don't underestimate the Bandersnatch," Pepper repeated ominously.

Lulu carefully reached into the bag and transferred 16 clusters of Tumtums, one cluster at a time into the basket. "We have 32 clusters, and we're giving you 16," she said, speaking slowly and annunciating each syllable in case the Bandersnatch was listening. "Half of 256 equals 128," she added to make sure the Bandersnatch understood.

"Let's cross," beckoned Elizabeth. Hand in hand, the twins walked timidly across the bridge. Pepper tromped clumsily behind them. Elizabeth was particularly annoyed by the noise that the Pig's hooves were making on the wooden boards.

Once they were safely across the bridge, both girls felt relief wash over them. When Lulu turned around, the basket was already empty.

"One crossing down, seven bridge crossings to go," she said.

The rest of the bridge crossings were similarly uneventful. In front of each bridge, they found the same type of wicker basket and the same type of sign. At the second bridge, the girls left 8 of their 16 remaining clusters. The backpack quickly grew lighter as the girls left half the remaining Tumtums at each bridge.

Lake Charles was a relatively small lake, and it wasn't long before the threesome was standing on Chataway Isle in front of the last bridge. This was the seventh bridge encountered, but eighth crossing since they had crossed twice over the bridge connecting the two islands.

"We had four Tumtums before the last bridge we crossed, and left two there," said Lulu, "so we should have two in our bag- one for the Bandersnatch and one for Mrs. Magpie. She reached her hand into the backpack to feel for the two remaining Tumtums. Lulu's fingers wrapped around only a single fruit.

"There's only one Tumtum in here," she spoke frantically. "Where did the other one go?"

"There were two left after the last bridge. I'm sure of it," Elizabeth said. She took the backpack from Lulu's hands, turned it upside down, and shook it. Some sheets of paper, pencils, a ruler, and a calculator scattered on the dirt. But no Tumtum fell out.

Both girls looked at Pepper. "I didn't take it," he insisted. "One large cabbage is enough for an afternoon snack."

"Well, what do we do now?" asked Lulu. Should we leave the one Tumtum in the basket? That's half of the two we were supposed to have left."

"But that's not half of the fruit we have. We have to follow the instructions precisely," Elizabeth insisted, "Or who knows what will happen," she cautioned.

"I can bite the Tumtum in half," offered Pepper, his mouth watering.

"No!" both girls yelled at once.

"I'll do it," Elizabeth said. "Hopefully, the Bandersnatch won't mind."

Lulu handed the fruit over to her sister, who wiped it vigorously on her shirt. Elizabeth turned the Tumtum around in her fingers trying to figure out exactly where the fruit should be cut to create two equal pieces. She carefully sunk her teeth into the fruit and took a slow, calculated bite. The Tumtum juice that dripped into her mouth was sweet. It reminded her of a cross between a ripe peach and a mango. She longed to take another bite of the delectable fruit but stopped herself.

Elizabeth was glad that she was the one who had volunteered to divide the fruit in half. She was now thoroughly convinced that neither Lulu nor Pepper possessed the self-control necessary for the job.

Elizabeth dropped the remaining half of the fruit into the basket. A low guttural growl rose up from beneath the bridge. The sisters exchanged terrified glances. "Run!" yelled Lulu. They both took off running, hearts beating wildly, not daring to look back.

They suddenly heard a loud thump. *Pepper tripped*, thought Lulu, *we need to help him*. She let go of her sister's hand and turned around. The Pig had indeed fallen. He was lying on the bridge, hooves in the air. A figure in a black cloak was standing over him. It glanced up at Lulu from beneath its dark hood and then vanished.

Lulu ran to Pepper's side. "Did the Bandersnatch do this to you?" she asked.

"Bandersnatch? No, no," blubbered Pepper, "You startled me with all that yelling and running, and I slipped. Pigs aren't meant to be bipedal, you know. Center of balance is too high, hooves are too narrow," his voice trailed off.

"Nobody's perfect," said Elizabeth, who had just joined the others at the bridge. Seeing Pepper in such a compromising position made her feel sorry for the fellow.

Pepper soon flipped himself over and regained his posture and composure. They continued walking along the trail.

Play Along

1. (2 pts.) Using your Tangram set (from Chapter 2), construct the following puzzle of a bridge:

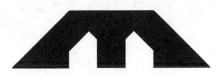

Fun fact: The ancient Romans built hundreds of stone arch bridges. Trajan's Bridge, the first bridge to be built over the Danube River, remained the longest arch bridge in the world for over 1,000 years.

2. (8 pts.) All the names on this map have something to do with Lewis Carroll. Do some research to figure out the relationships:

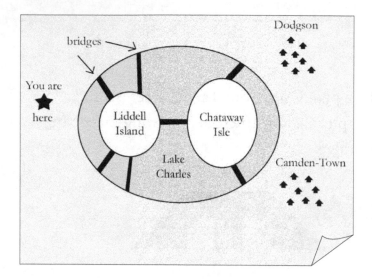

2. (3 pts.) With the Seven Bridges of Königsberg problem, Euler (pronounced "Oiler") started the field of graph theory. He redrew the map as a series of dots (called "nodes") representing land connected by lines (called "edges") representing the bridges. Can you model the map above in this manner? Label the node for the mainland "M", the node for Liddell Island "L", and the one for Chataway Isle "C".

3. (2 pts.) On the map you drew in the last problem, add a bridge that would allow you

to start at node "M", cross over each bridge once, returning to "M".

4. (4 pts.) Remember that each Bandersnatch will take half your fruit, and you want one fruit to remain at the end. How many Tumtums should you bring if you need to cross: three bridges? Five bridges? Nine bridges? Fifteen bridges?

5. (6 pts.) Test out Thales's method for calculating the height of an object. On a sunny day, place a stick into the ground. Measure the stick's height and then the length of its shadow. Keep checking the length of the stick's shadow throughout the day until the shadow is the same length as the stick (or half, or double the height of the stick for easy calculations). Then have a friend measure the length of your shadow to try to figure out your height. Calculate the height of a tall object like a tree, house, or telephone pole by measuring its shadow.

6. (3 pts.) Lulu and Elizabeth made a word ladder with twelve rungs from MATH to GAME. To form each new word, change one letter and keep the rest the same. This game was invented by Lewis Carroll and is sometimes called Doublets. Can you figure out the rest of their word ladder?

MATH

PATH

PITH

WITH

_____(blow out the candles and ___)

_____(synonym for clever, smart)

_____(to get up)

_____(a grain popular in Asia)

_____(a competition)

_____(a ratio or proportion)

GATE

GAME

7. (5 pts.) Find word ladders to transform:

   a. CAT to DOG

   b. FOOD to GOLD

   c. SEED to TREE

8. Veracity

Lulu's head filled with questions about the Bandersnatch. But before she could query the Pig further, Elizabeth stopped abruptly, causing Lulu and Pepper to nearly run into her. "Where exactly are we going?" Elizabeth asked. "To Camden-Town or to Dodgson?"

Lulu shrugged, and Pepper sneezed. Elizabeth opened Mrs. Magpie's book, which she had been carrying the entire time. She read:

> *Your journey is ending and it's time to come down,*
> *To either Dodgson or Camden-Town.*
> *Mome Raths reside within one pleasant village,*
> *The other succumbed to the Slithy Toves' pillage.*
> *To find the right place will take sagacity,*
> *And an uncanny ability to test the veracity*
> *Of the characters which on your way you may pass.*
> *Determine the truth, and we'll meet at last.*

Lulu instantly forgot all about the Bandersnatch. Her heart skipped a beat at the thought of finally meeting Mrs. Magpie. The feeling soon turned bittersweet, however, at the realization that their journey might soon be concluding.

"How do we figure out which town to go to?" asked Elizabeth. "The clue mentioned 'veracity' which means truthfulness. Maybe that has something to do with it," she surmised.

"It sure does," Pepper Pig interjected, stepping toward the girls with a smile. Elizabeth glared at him. Just because she felt sorry for Pepper didn't mean that her reservations about his character had vanished.

"I know you don't trust me after the umm… cabbage incident," he continued, addressing Elizabeth directly, "but I don't think you can figure out this clue without my assistance."

"Alright," Elizabeth relented, "what do we have to do?"

"The Mome Raths can be trusted because they always tell the truth. The Slithy Toves are wicked creatures with nefarious intentions. They never tell the truth," explained Pepper.

"Never?" asked Lulu

"Never. Slithy Toves will always lie, and Mome Raths will always tell the truth. And that is a fact."

"Then we just need to find a Mome Rath and ask him in which of the two towns Mrs.

Magpie can be found. He'll tell us the truth," concluded Lulu.

"Easier said than done," responded Pepper. "The Mome Raths and Slithy Toves belong to the same tribe. Telling them apart is tough. Only someone born and raised here, like me, can tell one from the other."

"And there's one more thing," he added. "Each Mome Rath and each Slithy Tove will only answer one question each day. So you have to ask carefully."

Lulu spotted some movement on the path between the two villages. Without further discussion, the three of them immediately headed in that direction. As they neared, Pepper motioned for them to stop.

"Look beside the Borogrove," the Pig whispered.

Lulu and Elizabeth looked down the path. Next to a small shrub was a miniature creature that resembled a guinea pig. It had one bright red flower in its fur and two more in its mouth, which it was busily munching.

"Which is it?" whispered back Lulu.

"A Mome Rath; I'm sure of it," said Pepper.

"How sure?" asked Elizabeth, head cocked and eyebrows raised.

"80 percent sure," said Pepper tentatively, "Maybe 75 percent."

"Nonsense!" a deep voice startled them from behind.

They all turned around to find themselves face to face with Manxome McLay, the Wolf.

"What you have there is definitely a Slithy Tove. I'm 95 percent sure of it," he said.

Lulu turned to her sister. "Who should we believe?" she whispered in her ear.

"I don't think it's smart to trust anyone except each other around here," Elizabeth whispered back.

"Trust me, then," said Lulu. She walked forward toward the creature. Before anyone could stop her, Lulu knelt down and spoke into the creature's ear. It dropped the flowers it was chewing and responded in a soft voice that only Lulu could hear before scampering behind the Borogrove. The Borogrove, which was not a shrub at all but a large bird, awoke and flew away with the small creature in its talons.

"Poor thing!" gasped Lulu. The Borogrove was already out of sight, and there was not much she could do. Lulu stood up and returned to the group. Manxome, Pepper and Elizabeth were all staring at her in disbelief. "I know which village we should go to!" Lulu announced cheerfully.

"Which one?" a perplexed Elizabeth asked.

"Dodgson," Lulu responded confidently.

"Are you sure?"

"100 percent!"

"Was the creature a Mome Rath or Slithy Tove?" inquired Pepper.

"I haven't a clue," said Lulu

"So how do you know if the creature was telling the truth?" It was Manxome's turn to ask.

Play Along: (4 pts.) What question did Lulu ask the creature to figure out which city was which? Keep reading for the answer.

Lulu, who was quite enjoying the questioning, decided to keep her secret. She started walking toward the town of Dodgson, quietly singing to herself:

> *"Twas brillig, and the slithy toves*
> *Did gyre and gimble in the wabe;*
> *All mimsy were the borogoves,*
> *And the mome raths outgrabe...[1]"*

Elizabeth matched Lulu's pace. Pepper and Manxome reluctantly followed behind them, staring disapprovingly at each other.

Soon they were standing in front of a sign that read "Welcome to Dodgson".

"Will you please tell me what you asked that creature?" Elizabeth pleaded. "I've been trying to figure it out and I can't. If you ask the creature which one was the Mome Raths' village and it was a Mome Rath, it would tell you the correct answer, but if it were a Slithy Tove, it would lie and lead you to the wrong town."

"It was simple," Lulu replied. "I knew that nobody had complete confidence about whether the creature was a truth-teller or liar, so I just asked which one was its home. If it were a Mome Rath," she continued, "it would tell me the correct village, and if it were a Slithy Tove-"

"-then it would lie and give you the name of the Mome Rath village rather than its own," jumped in Elizabeth. "That way the answer would always be the same- the town we need to reach. Clever!"

Manxome and Pepper, who were listening to the explanation, nodded in approval. Lulu beamed.

They looked around Dodgson. Lulu and Elizabeth had expected to see a village full of tiny houses for the guinea-pig-sized Mome Raths. Instead, a beautiful cobblestone path led past houses of different shapes and sizes, all with lavish gardens. Animals of every type (and some which the girls did not recognize) were going about their daily business and appeared to be oblivious to the group of strangers.

"How do we find Mrs. Magpie's house?" asked Elizabeth. She opened The Manual only to find that the next page was empty.

"Maybe we should ask someone," suggested Lulu. "We can trust the Mome Raths who live here." She knelt down in front of a small fellow who nearly walked into her. "Do you know where Mrs. Magpie is?" Lulu asked the startled creature.

"Yes," it replied in a small voice, and then kept walking on its way, taking a detour around Lulu's legs.

"We need to be more specific," Lulu concluded. "Since we can only ask each Mome Rath one question, it has to be the right one.

As she was speaking, another Mome Rath ran into her left ankle. "Where might Mrs. Magpie be?" Lulu quickly asked.

"She might be anywhere," was the reply before the Mome Rath scampered off.

"Don't say 'might,'" suggested Elizabeth.

Lulu did not have to wait long before a third Mome Rath crossed her path. "Where is Mrs. Magpie located right now?" she inquired.

The Mome Rath pointed to a large house across the street. Later, after thinking it over,

Lulu felt lucky to have received an answer at all since 'right now' is a relative term.

Four pairs of eyes turned toward the house. Turrets at its corners and flowering ivy framing a hand-carved wooden door gave it a castle-like appearance. A thin wisp of smoke appeared to be defying the breeze by rising vertically out of the stone chimney.

"Onward!" said Lulu, jovially.

As the group stepped toward the wooden door of the chateau, it swung open silently. Lulu and Elizabeth looked at each other, then both stepped inside.

They found themselves in a dim chilly hall. "I wonder if anyone lives here at all," whispered Lulu to her sister as she ran her finger across a short dusty table. The table held two brass candlesticks and an empty pewter bowl. These were all covered with a fine layer of soot. Lulu wiped her dusty finger on her skirt.

"Don't touch anything," Elizabeth hissed. She motioned for them to walk toward a connecting room, from which they could hear a hum of voices rising and falling.

Lulu, Elizabeth, Pepper, and Manxome walked to the doorway of this room but dared not enter. The room was packed with all sorts of animals, all speaking at the same time. A large horse stood right in front of them so they could not see what was going on.

"Silence, please," spoke a voice that was both firm and kind. All conversation suddenly ceased.

Pepper sneezed loudly. With the shuffle of hooves and wings and claws, all the animals turned at once to face the strangers. Elizabeth's cheeks felt warm with embarrassment. She could hear Lulu whispering Pi under her breath.

The horse moved aside and revealed a golden chair, studded with glimmering jewels. In sharp contrast to the dusty furniture in the previous room, this chair was impeccably clean. Lulu stopped reciting Pi. "A throne!" she gasped.

Perched on one of its arms was a bird with shiny white and glossy bluish-black feathers, not one out of place. A string of pearls adorned its throat. The bird was so still that at first Lulu thought it was part of the throne's decor. Then it, too, turned its head toward the strangers.

"Welcome, Lulu and Elizabeth" spoke the bird.

"Mrs. Magpie!" Lulu blurted out.

"We're very pleased to meet you," spoke Elizabeth in a small voice, adding a little curtsy.

"Formalities are unnecessary," said Mrs. Magpie, "I have been watching your journey. You've proved yourselves to be ingenious young ladies. You make an excellent team."

Lulu and Elizabeth looked at each other and smiled.

"Come forward, please," she commanded.

The twins began walking toward her. Pepper and Manxome followed.

"Only the ladies, please," reproached Mrs. Magpie. Pepper opened his mouth to protest but then closed it, giving him the appearance of a codfish.

"I require your assistance," Mrs. Magpie explained. "You may already be aware that I am the acting Regent of Tulgey Wood until the Council appoints a proper ruler. One of my responsibilities is to resolve disputes to maintain order and justice in these parts. I would like you two ladies to aid me in this task today."

"We'd love to!" Lulu cried out. She had often dreamed of sitting on a fancy throne and ruling a kingdom.

"Sure," Elizabeth agreed timidly. She was pleased that they'd finally be able to use their brains to help people.

"Please have a seat," Mrs. Magpie motioned to the throne. Lulu could hardly contain her excitement. Both girls fit in the opulently decorated chair with room to spare.

A Kangaroo wearing a flowered hat stepped forward. At her feet were a dozen very sulky-looking Mome Raths.

"Mimsy Kangaroo, you may proceed," spoke Mrs. Magpie.

The Kangaroo wasted no time. "I hired these Mome Raths to clean my house. You know how you can always trust Mome Raths," she addressed Mrs. Magpie. "But one of them is an impostor and has stolen from me," the Kangaroo continued. The other animals in the room began to whisper among themselves.

"Silence," spoke Mrs. Magpie again.

"When I went to the grocer's this morning, Mr. Duckbill refused to take my coins. He said that when he put them in the coin verification machine, it told him that one of them was a fake. There must be a Slithy Tove posing as a Mome Rath in this group," the Kangaroo pointed accusingly at the quivering Mome Raths at her feet.

"First, we must make sure that one of the coins is indeed a fake," said Mrs. Magpie. "There is no use accusing an innocent creature if no crime has been committed."

Lulu marveled at Mrs. Magpie's composure and rationality. She, herself, had already been running through what questions she'd ask the Mome Raths to figure out which one was a Slithy Tove.

"How does the coin verification machine work?" asked Mrs. Magpie.

The Kangaroo wrung her fingers nervously. "Well, I don't quite know," she admitted. She produced a leather satchel and emptied it into her palm to reveal nine coins. "They all look identical to me," she said.

"This is just like the story about Archimedes," Elizabeth whispered to her sister.

"Speak up, my dear," encouraged Mrs. Magpie.

"The king asked Archimedes to find out if his crown, which he had commissioned out of solid gold, actually had silver mixed in," explained Elizabeth. "The crown weighed the same as the gold brick the king had given the goldsmith. However, the king wasn't quite sure if the goldsmith had kept some of the gold for himself and made the crown by mixing some gold with less expensive silver," Elizabeth continued. She suddenly remembered all the eyes that were staring at her and began to blush.

"Go on, darling," urged Mrs. Magpie, "tell us Archimedes's solution."

"The solution was to put the crown into a container of water and see if it displaces the same amount as the brick of solid gold," Elizabeth said.

"You forgot the best part," interrupted Lulu. "He came up with the solution while he was in the bath," she said, "and ran through the streets naked shouting 'Eureka!'" The crowd burst into laughter, and Elizabeth's cheeks flushed redder.

"You may use my cup to test the coins," offered Mrs. Magpie.

Elizabeth noticed the side table for the first time. It contained a calculator, a balance scale, a quill pen, and a glass of water. She hesitated.

"Wait," Elizabeth spoke softly. "Before comparing the volume of each coin, we should check to see if they are really all the same weight. It will take four weighings to compare the nine coins and find out which one is fake."

"Nonsense," a voice in the audience called out. Manxome McLay stepped forward. "I can figure it out with three weighings," he spoke slowly and deliberately. "Maybe I'm the one who deserves to be up there." He motioned to the throne.

Elizabeth felt faint. She didn't know what to say and looked wide-eyed at Lulu in desperation. Lulu stood up. "I can do it in two," she announced.

Manxome's laugh echoed through the room. "Just you try," he chuckled.

Lulu reached for the nine coins. "May I?" she asked. The Kangaroo nodded.

Play Along: (4 pts.) Assume that out of the nine coins, one is lighter than the rest. How would you figure out which of the coins is lighter using the balance scale only twice? Draw a balance scale on a piece of paper (or just two rectangles to represent the two sides). Take nine small objects, such as beans, paperclips, or crackers to represent the coins, and demonstrate how you would compare their weights. Keep reading for the answer.

Lulu approached the balance scale and put three coins in each of its two buckets. She held the three remaining coins in her hand. The scale balanced perfectly.

"If there is one coin that does not weigh the same as the rest, it is not among these six coins," she declared.

Lulu removed the six coins from the scale and set them aside. She opened her hand to reveal the remaining three coins. Lulu placed one coin on each side of the scale, leaving one in her hand. The scale tipped to the right.

"The coin on the left weighs less than the coin on the right," she announced, "so it must be the fake coin." She was glad that the coins hadn't balanced. She had just remembered that they didn't know whether the fake coin weighed the same as the others. If the coins had balanced, then she would have no way of knowing whether the coin in her hand would have been a different weight. Lulu decided to keep that realization to herself and not admit her mistake.

"Not so fast!" snickered Manxome. "You don't know for sure that the counterfeit coin weighs less than the rest. What if it weighs more? It could very well be the coin on the right," he said.

Now it was Lulu's turn to blush. She replaced the coin on the right with the one in her hand. The scale tipped to the right once again.

"You weighed the coins three times. You've failed," claimed Manxome.

"But I was correct – the coin on the left was the different one," insisted Lulu.

"There is no shame in making mistakes," said Mrs. Magpie kindly. "That is how we learn not to jump to conclusions too hastily."

Lulu knew that Mrs. Magpie was right, but she still felt deflated. She hung her head as she took her seat. Elizabeth put a consoling hand on her sister's shoulder. "It's alright. You did great," she whispered in her ear.

"Let's continue, then," began Mrs. Magpie.

"Excuse me, please," interrupted Elizabeth, "but I do believe Manxome was also mistaken."

Lulu stared at her sister in disbelief. *What was she doing?*

"Now, we know the fake coin weighs less than the rest, but we didn't know that before. So assume that we don't know whether the counterfeit coin weighs more, less, or the same as the other eight coins. Using the balance scale three times, as Mister Maclay suggested, is not enough to guarantee that we will find the answer," said Elizabeth, her voice growing stronger as she became more sure of herself.

"Let's say that we put three coins on each side of the scale, and it tipped to the right. We can only eliminate the possibility that the fake coin weighs the same as the rest. But we can't tell whether it is in the group on the right, or on the left. We need to weigh the coins in one of the groups against each other. And even then, they may all be equal, which means we would have to weigh the coins in the other group against each other. That's a lot more than three weighings," she said with a smile. The smug look left Manxome's face.

Mrs. Magpie was impressed, but she also had a job to do. "I sincerely appreciate the clarification, Elizabeth," she said. "Lulu and

Manxome were both mistaken, but this is neither a competition of brains nor egos. She glared with her sharp black eyes first at Manxome and then at Lulu. "We are in pursuit of justice. Let us carry on."

Play Along

1. (2 pts.) Using your Tangram set, construct the following puzzle of Mimsy Kangaroo:

Fun fact: You wouldn't have found this kangaroo puzzle in ancient China where Tangrams reportedly originated; Kangaroos only live in Australia. A red kangaroo can cover a distance of 25 feet (7.6 meters) in a single hop.

2. (1 pt.) Mome Raths always tell the truth, and Slithy Toves always lie. If you meet a creature and ask if it is a Mome Rath, what would it say?

3. (3 pts.) You meet two creatures, Alex and Betty. Alex says, "We are both Mome Raths." Betty says, "Don't believe his lies." Is Alex a Mome Rath or Slithy Tove? What about Betty?

4. (5 pts.) You meet three creatures, Ann, Bob, and Carl. Ann whispers something to Bob. "She told me that she is a Slithy Tove," says Bob. "Ann and I are both Slithy Toves," says Carl. Which of these creatures are Mome Raths and which are Slithy Toves?

5. (3 pts.) Have some fun experimenting with Archimedes's buoyancy principle. Take a clear container of water and mark the water level with a dry-erase marker. Place various objects into the container and mark the new water level. Use a scale to weigh the objects. Can you find an object that weighs more than another but displaces less water?

6. (3 pts.) Assume you have three coins and knew that one coin was lighter than the other two.

What is the smallest number of times you have to use the balance scale *to be certain* of which one is the counterfeit coin?

7. (3 pts.) What is the minimum number of times that you would have to use the scale in the previous problem *to be certain* of which of the three coins is counterfeit if you didn't know whether the counterfeit coin weighed more or less than the other two?

8. (6 pts.) Consider the problem with the nine coins where you don't know whether the fake coin weighs more, less, or the same as the other eight. Lulu and Manxome were both wrong, so it's your turn to try. Think of the worst case scenario. How many times would you have to use the balance scale before you get your answer?

9. To Catch a Thief

Mimsy Kangaroo stepped forward. "Thank you for finding the counterfeit coin," she said. "But there is still the matter of which one of the workers is an impostor."

The Kangaroo turned to face the terrified creatures. "Which of you is a Slithy Tove?" she asked angrily.

Twelve little voices spoke in unison, "Not I!"

Elizabeth and Lulu were both thinking that the Kangaroo was treating these critters very unkindly, and felt sorry for them.

"Mrs. Magpie," Lulu asked in a soft voice, "when we discover the Slithy Tove, what will happen to him or her?"

Mrs. Magpie replied, "Well he'll have to return the stolen coin, of course. Then he will have the choice of joining the other Slithy Toves in Camden-Town, or going through rehabilitation."

"Rehabilitation?" asked Lulu.

"Yes, Slithy Toves are born as Mome Raths. Once they consciously choose to tell a lie, they can no longer tell the truth. In Tulgey Wood, you are in control of your destiny, whether good or bad. Reform is not easy, but it is possible. There are a couple Mome Raths in our community who had been Slithy Toves at one point in their lives."

Lulu breathed a sigh of relief. She was glad that the thief would not come to any harm.

"I think I can help find the Slithy Tove among these Mome Raths," she said.

"Remember that they have already been asked a question and will answer no others today," Mrs. Magpie reminded her.

"I know," said Lulu, standing up once again.

"I am a great sorceress," she spoke in a dramatic tone, "and I have enchanted the candlesticks in the other room to reveal the identity of the Slithy Tove to me."

"Are you crazy?" whispered Elizabeth, staring at her sister in disbelief.

Lulu ignored her and continued. "Each of the suspects must rub the candlesticks to activate the magic."

One by one, each creature left for the other room to rub the candlesticks and returned solemnly to its place in line. Magic was common in Tulgey Wood, so Lulu's request was not unusual.

When all the Mome Raths had taken a turn in the other room, Mrs. Magpie asked the Kangaroo to fetch the candlesticks for Lulu.

When Elizabeth saw the candlesticks and remembered how dusty they had been, she instantly understood her sister's plan.

Lulu walked along the line of Mome Raths, waving the candlesticks and examining each creature carefully. She stopped in front of the last creature who had his hands tucked in his fur. "Please show me your hands," she requested. He had pulled out one hand and was about to reveal the other when Lulu announced, "This one is a Slithy Tove."

The creature pulled his other hand from his fur and a coin fell onto the floor with a clink. Lulu had uncovered the thief. Mrs. Magpie let out a sharp whistle and a Heron promptly appeared. He picked up the Slithy Tove in his beak and whisked him out of the room.

Lulu returned to the chair, quite pleased with herself.

"John Napier's rooster?" asked Elizabeth.

Lulu nodded. John Napier was a mathematician that the girls had read about recently. Besides being known as the inventor of logarithms and a calculation device called

Napier's bones, he was rumored to be a sorcerer. One of the stories about John Napier is that he found out which of his servants was stealing from him by claiming that he had a magic rooster that would identify the thief.

He asked each of the servants to go into a dark room and pet the magic rooster. The servants who had nothing to hide followed the instructions, but the thief, believing the rooster was magical, would enter the room but not pet the rooster. Napier had rubbed black soot on his rooster, so he could identify the thief by his clean hands. Lulu had used the same trick to find the Slithy Tove.

"I thank you from the bottom of my heart," said Mimsy Kangaroo. "You have no idea how much hassle this saves me. It's not just about a single fake coin. I need workers I can trust."

"I'm happy to help," said Lulu.

"But I want to reward you," insisted the Kangaroo. "What currency do you have in the land you come from?" she asked.

175

"A dollar in their land is worth about two Dooplas in ours," said Mrs. Magpie.

The Kangaroo nodded at Mrs. Magpie, then turned back to face Lulu. "I'll give you a choice," she said. "Would you like to receive ten dollars every day for twenty days? Or would you prefer to get five dollars a day for the first fifteen days, and twenty-five dollars a day for the final five days?"

Play Along: (2 pts.) Consider Mimsy Kangaroo's two offers. Which one should Lulu choose to receive the largest reward? (Keep reading for the answer – cover the next page while you work).

"Both offers are the same - two hundred dollars," Elizabeth whispered.

Lulu thought for a second. "I would like to receive payment in pennies," she announced.

"Pennies?" asked the Kangaroo, turning to Mrs. Magpie for clarification.

"One dollar is equivalent to one hundred pennies," explained Mrs. Magpie.

The Kangaroo grinned. There was no doubt that she thought Lulu was a very silly girl.

"I'd like one penny on the first day and double the pennies on each subsequent day for twenty days," Lulu requested politely.

The Kangaroo did a quick calculation in her head. *One on the first day, two on the second day, four on the third day, eight on the fourth, and sixteen on the fifth,* she thought, *which makes only thirty-one cents in the first five days. Not even a third of a dollar when she could have had fifty dollars in her pocket by that time. This child is quite foolish indeed!*

The Kangaroo's grin widened. "I accept the offer," she said.

"And one more thing," added Lulu, "I want to donate the reward money to the Slithy

Tove rehabilitation program. Everyone deserves a second chance."

Elizabeth stared at her sister in astonishment. *Was this the same girl who would talk endlessly about the elaborate costumes she would buy if she were wealthy?*

Lulu seemed to read her mind. "It just felt right," she whispered.

Mrs. Magpie flew over to the calculator and with five pecks on its keys had an answer. She then stepped over to the quill pen, picked it up with a claw, adroitly dipped it in ink, and wrote the sum on a small scrap of paper.

The Kangaroo stepped up to the small table, took one look at the figure, and with a shriek, fell to the floor, knocking down the water. Some animals immediately rushed to help her back to her feet. As they escorted her away, she glared at Lulu. "Well, I never!" she said on her way out.

"What was the total?" asked Lulu.

Mrs. Magpie pushed the scrap of paper in her direction. It was now wet on one side. Lulu looked at it. The figure was $10,485.75. *Those pennies sure add up*, she thought.

"Mrs. Magpie," Lulu said, "Do you think it was wrong to take advantage of the Kangaroo's generosity?"

Mrs. Magpie turned her head, so her dark eyes were looking right into Lulu's. "Mimsy Kangaroo could have done the calculations herself. Those who don't take the time to think completely through offers can be easily tricked," she responded.

"I think I'll only ask that she double the pennies for fifteen days instead of twenty," Lulu said, reconsidering. "That would be fair."

Mrs. Magpie nodded. "We will take a short recess before the next case," she announced. Some of the animals stepped outside. Others lingered in the room, chatting among themselves.

"Did the Bandersnatch tribute give you the idea of doubling pennies?" asked Elizabeth.

"Actually, it was the story about the man who invented the game of chess that inspired me. After this man had shown the king how to play

179

the game, the king offered to give him anything in the kingdom. The man only wanted to be paid in wheat – one grain on the first square of the chessboard, two on the second, four on the third, and so on…"

"For all sixty-four squares on the board?" asked Elizabeth.

"Yep. Can you imagine how much wheat that was? The king soon discovered there wasn't enough wheat in the entire kingdom to make the payment," said Lulu, smiling at the thought.

"I'm sure glad Mother encourages us to read stories of famous mathematicians. As Newton would say, *'If I have seen further it is by standing on ye sholders of Giants.'*[4]" quoted Elizabeth.

"I have another Isaac Newton quote for you," said Lulu, "*'Amicus Plato - amicus Aristoteles - magis amica veritas*[4]'"

"I see Latin lessons are going well. Translation, please?" said Elizabeth.

"*'Plato is my friend - Aristotle is my friend - but my greatest friend is truth.'*"

"Time for soup!" Mrs. Magpie interrupted. Lulu and Elizabeth suddenly noticed their rumbling stomachs. *They hadn't eaten a thing since breakfast!*

A Tiger moved a long table in front of the throne, and slapped down three placemats, all in a row - a red one in front of Mrs. Magpie, a pink one for Lulu, who was delighted to have received her favorite color, and a green placemat for Elizabeth.

The Tiger returned and stood in front of Elizabeth, who squirmed in the chair uneasily at the feel of his warm breath on her face. He dropped a stack of three bowls on the green placemat with a clank and walked away. Elizabeth noted that the bowls were of three different sizes. A small bowl was nested inside a medium sized bowl, which in turn was sitting inside a larger bowl.

"He did that improperly," huffed Mrs. Magpie. "The soup cannot be served until all the bowls are on the red placemat. Elizabeth grabbed the three bowls and was about to pass them over when Mrs. Magpie gasped. "No, no," she screeched out, "only one bowl can be moved at a time." Elizabeth could not make much sense of this instruction, but quickly pulled her hands away.

"They must be magic," Lulu whispered to her sister, comfortingly. She picked up the smallest bowl and moved it over to Mrs. Magpie's place-mat.

Lulu then grabbed the medium sized bowl and tried to put it on top of the smaller one, but before she could put it down, another whoop came from Mrs. Magpie. "You must always stack the bowls in the right size order! A bowl may never rest on top of a smaller one!"

Lulu quickly put the medium-sized bowl on the pink mat. *Now where could the large bowl go?*

Play Along: (5 pts.) On a separate piece of paper, draw and cut out three circles - small, medium, and large - to represent the bowls. On another piece of paper, draw the three placemats. Stack the bowls on the first mat, and transfer them to the 3rd mat, one at a time, remembering that a bowl can never be put on top of a smaller bowl. This is a classic logic puzzle called The Towers of Hanoi. Keep reading for the answer.

Elizabeth jumped in. "This is a lot like the river crossing," she said, taking the smallest bowl off the red mat and placing in on top of the medium-sized bowl. Lulu caught on. She then moved the largest bowl from the green mat to the red one. Since Elizabeth's placemat was empty, she moved the smallest bowl there. Now the medium bowl could go on top of the large one, followed by the small bowl.

The instant that the three bowls were stacked in the proper order in front of Mrs. Magpie, the smallest bowl began filling with soup. *"Beau-ootiful Soo-oop!*₁" Lulu crooned.

"Pass that one to Elizabeth," Mrs. Magpie instructed. Lulu, salivating, carefully picked up the steaming bowl and placed it in front of her sister, who sat patiently waiting for a spoon. The medium-sized bowl had now filled itself with soup. With a wink from Mrs. Magpie, Lulu took it in both hands and began sipping the delicious broth.

"Mock turtle soup- the best in all of Wonderland!" Mrs. Magpie raved before plunging her beak into the largest bowl. Lulu, her mouth full, nearly choked.

"I guess this is your 'big bowl of yum'," whispered Elizabeth, wrinkling her nose and pushing her own bowl away.

Lulu mulled over the idea of letting Elizabeth know that mock turtle soup actually doesn't contain any turtle meat, only a calf's head and various animal organs, but decided against it. She shrugged and put the bowl back to her lips.

Play Along

1. (2 pts.) Using your Tangram set, construct the following puzzle of the candlestick that uncovered the thief:

Fun fact: The anecdote about the thief-finding rooster was not the only colorful story about the mathematician John Napier. Another story tells of a neighbor's pigeons that infuriated Napier by eating seeds and grains from his field. Napier told the neighbor that if he ever catches those pigeons he is going to keep them. The neighbor agreed, thinking there was no way that the birds could be caught. The next week, the neighbor was shocked to see Napier picking the pigeons up off the field and putting them into sacks. How did he do it? He soaked dried peas in brandy (a type of alcohol) and scattered them in his field. The pigeons were easy to pick up because they were all intoxicated (drunk).

2. (2 pts.) Gather a collection of small objects (small crackers, cereal, buttons, beads, etc.). Start with one object on the first day, then double it for the second day, and so on. On a separate paper, make a table with columns for "Day" "Number" and "Sum". Keep going until you run out of objects. Take a look at the running sum and see if you can find a fast way to calculate the sum for any given day without adding up all the numbers. Don't read the next question until you have found the pattern.

3. (3 pts.) How did Mrs. Magpie figure out the amount owed for 20 days of doubling pennies by only pecking five buttons on the

calculator? You may have figured out the pattern in the last question. The sum is equal to the number owed on the next day minus one. You can calculate this by using an exponent. Ask a grown-up for a calculator and look for a button that looks like this: x^y Here are the buttons that Mrs. Magpie pressed (try them on your calculator):

      -First: The button to turn the calculator on (usually "C" or "ON)
      -Second: 2
      -Third: x^y
      -Fourth: 20
      -Fifth: = (some calculators don't require this)

Finally, Mrs. Magpie subtracted one from the result in her head and divided by 100 (since there are 100 pennies in a dollar) by moving the decimal point two places. You can use the calculator to do this if you'd like.

4. (4 pts.) Lulu changed her mind and only requested that Mrs. Kangaroo double the pennies for fifteen days instead of twenty. Use your calculator and the method you learned in the last problem to figure out the new amount she'll be paying.

5. (4 pts.) The stacking soup bowls were a variation of a game called The Towers of Hanoi. Legend tells of a monastery in Hanoi, Vietnam, that's contains three tall metal pillars. 64 golden disks of gradually decreasing size were put on the first pillar in the beginning of the world. The monks have to stack all the disks, in order of size onto the third pillar, moving only one disk per day, and following the rule that a larger disk cannot be placed on top of a smaller one. When the monks complete this task, legend says, the world will end.

Make your own 5-disk Towers of Hanoi game using paper (start with the circles and place-mats you already made in this chapter and add two more circles). Count the number of moves that it takes to solve the puzzle using 2, 3, 4, and 5 disks, respectively. Does the pattern look familiar?

6. (5 pts.) Use a calculator and the technique for calculating the doubling pennies to compute the number of days it will take the monks to finish solving the Towers of Hanoi puzzle with 64 disks.

10. The Vorpal Blade

The recess was soon over, and the animals filed back into the small room. The long table was removed, and twelve chairs were set up in a half-circle in front of Mrs. Magpie. A dozen animals of different kinds sat down in the chairs.

Mrs. Magpie went straight to business. "Members of the Tulgey Wood Council, I have a proposition for you today," she said, "As you all are aware, my position as Regent is, and has always been, temporary. Our search for an intelligent, fair, and selfless ruler has been long and arduous. But today I believe we've found two worthy candidates: Lulu and Elizabeth Lovelace." She pointed at the twins with a shiny black wing. Elizabeth shifted uneasily in her seat while Lulu radiated pride.

"I propose that we bestow these two fine young ladies with the Vorpal Blade." The animals in the room began speaking at once. Mrs. Magpie's voice grew loud and shrill to overcome the cacophony. "Silence!" Elizabeth thought that this was the first time that she sounded like a magpie.

The noise in the room dampened to low mumble. "It would be on a provisional basis, of course," Mrs. Magpie continued addressing The Council, her voice calmer, "until, under my mentorship, they have proved themselves capable of representing the values that embody the Vorpal Blade. Please let me know when you've come to a unanimous decision."

Lulu and Elizabeth looked at each other. The animals in the Council were discussing the matter among themselves with bowed heads and hushed voices.

Elizabeth turned to Mrs. Magpie. "What does this all mean?" she asked.

"The Vorpal Blade was used by the first Prince of Tulgey Wood to defeat the first Jabberwocky," Mrs. Magpie said.

Elizabeth and Lulu waited for her to continue. Only an awkward silence ensued.

"If this sword becomes ours, will we have to go around slaying monsters?" Lulu asked hopefully.

"Heavens, no," cackled Mrs. Magpie, "the Vorpal Sword is purely symbolic nowadays. It will make you Queens of Tulgey Wood, the finest province in Wonderland."

A Queen, thought Lulu, *is even more important than a countess!* Before Mrs. Magpie could explain just what being a ruler of Tulgey Wood entailed, the conversation was interrupted by a long succession of sneezes. Pepper Pig and Manxome Maclay were standing in front of Mrs. Magpie.

Manxome cleared his throat. "With all due respect, Your Honor," he began, addressing Mrs. Magpie directly in his deep voice. "These girls would never have reached you without our assistance."

"Indeed," added Pepper between sneezes.

"Credit shall be given where credit is due," said Mrs. Magpie evenly.

Just then, one of the Council members stood up. It was a large Walrus wearing a vest. "We've come to a decision," he announced. The room grew quiet, and all eyes turned to face him.

Before the Walrus could continue, a gust of wind burst into the room like a small cyclone. Papers swirled wildly about, and the wind blew Manxome's derby hat right off his head. At the window (which Mrs. Magpie had opened during the recess), was a grotesque face, shrouded by a hooded cloak. It had the beak of a raven, the jaw of a hyena, and eyes that held a terrifying almost-human quality to them.

"Bandersnatch, what is your business here?" asked Mrs. Magpie sharply.

The creature's bill protruded into the room, and he dropped something onto the floor. With horror, Elizabeth recognized it as the half Tumtum fruit that she had left at the last bridge.

A Mome Rath quickly scurried across the room and grabbed the fruit, consuming it as he ran away.

"This is unacceptable! The strangers have breached our contract," the Bandersnatch spoke with a refined musical voice that took Lulu by surprise. She did not believe the voice, which resembled the melancholy notes of a cello, could be coming from the monstrous face at the window. *Was this the same creature that had produced the raspy snarl at the bridge?*

"But we had one Tumtum in our possession and left half of it for you. That's what the sign said to do," Lulu spoke up. Elizabeth was amazed at her sister's bravery. She, herself, would never have the courage to speak directly to the ugly beast.

"On the contrary, I do believe the sign referred to dividing your Tumtum *stash*," objected the Bandersnatch, "'Stash' is defined, in the Wonderland Dictionary, as 'an amount of something that is concealed'. An 'amount' is defined as 'the aggregate of two or more quantities' - a sum, for those here unfamiliar with three-syllable words. Case closed."

Lulu thought the Bandersnatch was rather arrogant, but she had to admit he was well-read. *Does this Wonderland Dictionary define all of Lewis Carroll's nonsense words?* she wondered.

"Or 'stash' could be short for mustache," chimed in Pepper. As the Pig leaned forward, a Tumtum fruit slid out of his pocket onto the floor. He smashed it against his face, just above his snout, so it vaguely resembled a small mustache. Juice was running down his face. *The missing Tumtum!* Lulu and Elizabeth both thought at once.

The animals in the room burst into laughter, perhaps relieved to have a comedic respite from the seriousness of the matter at hand. The Bandersnatch's eyes narrowed with a look of disdain. He continued speaking.

"What if someone decides to carry a single Tumtum and leave half at each bridge crossing. By the 8th crossing, only $1/256^{th}$ of a fruit will be left. How can one expect to feed a family on that?" said the Bandersnatch, tears in his eyes.

Lulu tried, with great difficulty, to imagine how a Bandersnatch's wife and children might look.

"Or what if the visitors are particularly intelligent and decide not to carry any fruit at all, leaving half of zero which is zero at every bridge," piped in Manxome.

"Math should be used to help, not trick. Case closed," repeated the Bandersnatch, this time with a tone of deep sadness in his voice.

Lulu waited to see if he would continue speaking. Evidently, the other animals were thinking the same, as there was a moment of silence, which only served to add dramatic flair to the Bandersnatch's words.

Lulu was half convinced by the Bandersnatch's passionate plea. Then she remembered what Pepper had said about the Bandersnatch gang asking for the tribute in exchange for not terrorizing the villagers. "I don't know your story," Lulu blurted out, "but where I come from, rewards must be earned. You are obviously not a stranger to the English language and could very well build a word ladder for yourself and harvest your own Tumtums.

When I am Queen, things will change around here."

The Bandersnatch looked confused. He turned to Mrs. Magpie. "To what is this child referring?" he asked.

"The Council was just deciding whether to bestow the Vorpal Blade onto these two young ladies," Mrs. Magpie explained.

"Preposterous! Absurd! Ludicrous! Irrational!" cried out the Bandersnatch.

"As irrational as the square root of two?" asked Manxome.

The Bandersnatch ignored him. "Outrageous! Unthinkable! Ridiculous! Asinine!" he continued.

Lulu thought the Bandersnatch was doing an excellent job of re-enforcing her point that he was capable of forming word ladders on his own, since he seemed to be a walking thesaurus.

"That will suffice!" Mrs. Magpie stopped him.

The Bandersnatch regained his composure. "Well, the Law allows any citizen to challenge candidates to prove whether they are suitable for the position. I would like to exercise that right."

"You do have that right," Mrs. Magpie agreed, "but I must tell you that I have already personally vetted these girls. They have encountered a plethora of math and logic problems on their journey here. All were solved accurately and tactfully."

"This test shan't take long," said the Bandersnatch. "It is only a simple riddle."

Mrs. Magpie gave Lulu and Elizabeth an encouraging look before turning back to the Bandersnatch. "You may proceed."

The Bandersnatch revealed a scroll, which he unrolled slowly, stopping frequently to scan the room and make sure that all eyes were on him. He cleared his throat and read:

Add the odds to get a square,
Maybe primes are not so rare,
One-one-two-three heeds nature's call
Reveal the thing that conquers all.

Elizabeth suddenly remembered that she was still clutching *Mrs. Magpie's Manual.* Even though the book's author was sitting right next to her, she instinctively felt that the manual could help. She quickly flipped through the pages to see if the book held a clue to the Bandersnatch's riddle.

"You have no more use for that book," said Mrs. Magpie, "as the remaining pages are empty. My writing was cut short by an unfortunate incident..."

"The Jabberwocky?" Elizabeth asked.

Mrs. Magpie nodded solemnly but did not elaborate further.

"Can my sister and I have some time to think it over?" Lulu asked.

"It's a simple riddle," repeated the Bandersnatch.

"You may," answered Mrs. Magpie.

Play Along: (15 pts.) Do you have what it takes to rule Tulgey Wood? Try to solve the Bandersnatch's riddle. (Keep reading for the answer.)

Lulu and Elizabeth huddled together, their heads touching. "I wrote the riddle down," said Elizabeth, unfolding a sheet of paper. Lulu was impressed by how proactive her sister had been.

"*Add the odds to get a square*," read Elizabeth, "I think I know this one." She started drawing on her paper.

"If you take one square, and add three more squares, you get a big square," said Elizabeth.

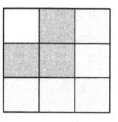

"Add five more, and you get a larger square," she continued.

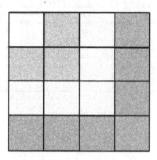

"So we added 1, 3, and 5. Add seven more squares and voila! Adding the odd numbers gets you a square."

"Not just a square," said Lulu, "but a square number. 2x2=4, 3x3=9, 4x4=16. I bet that adding seven more will get you a large square that is 5 by 5, or 25." She smiled to herself, thinking about how Gauss's formula for calculating the sum of consecutive numbers would work for just the odd ones.

Elizabeth looked a lot more serious; she didn't want to get distracted from the important problem at hand. *"Maybe primes are not so rare,"* she read the next part of the clue.

"A prime number can only be divided by one and itself," said Lulu.

"Yes, like five or seven," added Elizabeth and began drawing on another piece of paper.

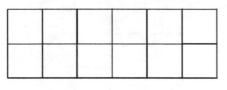

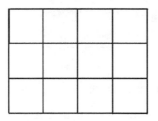

"So let's say you have twelve blocks," said Elizabeth, "you can arrange them into all sorts of rectangles... 2 by 6, 6 by 2, 3 by 4, 4 by 3, 1 by 12, 12 by 1."

"Right," said Lulu, "twelve is not a prime number."

"But take a prime number like 7, and the only way you can arrange those blocks to form a rectangle is in a long straight line. 7 by 1, or 1 by 7."

"But what about the clue," asked Lulu, "it's true that prime numbers are not so rare. There is an infinite number of them. Mathematicians are using computers to find new ones all the time.

"Let's keep going," suggested Elizabeth. *"One-one-two-three heeds nature's call."* she read.

"That's the Fibonacci sequence," Lulu noticed, "just add the last two numbers to get the next. 1+1=2, 2+1=3, 2+3=5." She wrote on Elizabeth's paper: 1, 1, 2, 3, 5, 8, 13.

"That pattern appears a lot in nature. It's in flower petals, pinccones, snail shells, and more," pointed out Elizabeth, "but we should be getting to the last line. *Reveal the thing that conquers all.* What does that mean?"

"Time is up!" yelled the Bandersnatch, "I need an answer. You may have one guess each."

Lulu and Elizabeth looked up with surprise. They had been so busy playing with numbers that they had forgotten about everyone else in the room.

Manxome and Pepper were still standing at the front of the room. Manxome turned to face Mrs. Magpie. "Are the human girls the only ones who can guess the solution to the riddle?" he asked.

"The Law allows anyone to accept the challenge, but each individual may only give one answer," replied Mrs. Magpie.

"Then I have one," said Manxome, "The answer is thirteen. It is an odd number, a prime, and also in Fibonacci's sequence. Thirteen is a part of all the sections of the clue, so it conquers all."

Lulu and Elizabeth looked at each other in astonishment. *How did Manxome know so much about math? And why hadn't they come up with this answer?*

"Wrong answer!" yelled the Bandersnatch, much to the relief of the crowd who did not fancy the idea of being ruled by someone of questionable character such as Manxome. "Anyone else?"

"42?" called out an anonymous voice in the room.

"Wrong book, Deep Thought!" snapped the Bandersnatch. Lulu, who had just finished reading *The Hitchhiker's Guide to the Galaxy*, suppressed a giggle.

"If you please, Madam Magpie," Pepper spoke up. "I have an answer, as well," he said and began to sneeze.

"Let's get on with it," said Mrs. Magpie impatiently.

"Infinity," said the Pig, "Infinity is larger than any other number, so it conquers all."

The animals in the room nodded to each other that this was a wonderful answer. Even Elizabeth was thinking that nothing can be greater than infinity.

The Bandersnatch spoke in a soft voice. Everyone strained to listen.

"Even as the finite encloses an infinite series
And in the unlimited limits appear,
So the soul of immensity dwells in minutia
And in narrowest limits no limits inhere.
What joy to discern the minute in infinity!
*The vast to perceive in the small, what divinity!*₅*"*

"He's quoting Jacob Bernoulli, the Swiss mathematician," whispered Elizabeth.

The Bandersnatch shook his head vigorously as if he were snapping himself out of a daze. The black hood fell at his shoulders revealing a head of dark feathers. "Even an infinite series can have limits. Your answer is incorrect!" he shouted dramatically. "But a good try," he mumbled under his breath.

Pepper was so pleased with himself that he couldn't stop sneezing and had to step outside for some fresh air. All eyes turned to Lulu and Elizabeth. Elizabeth froze in panic, but when her sister's hand wrapped around her arm, she felt a sudden surge of confidence. Since the Bandersnatch seemed somewhat happy with Pepper's answer, she figured that he was on the right track. *What could be so vast that it even includes infinity?* she thought.

"Mathematics," Elizabeth blurted out. "The entire universe can be reduced to mathematical equations. Math can solve any problem. It's the greatest power of all."

The Bandersnatch inhaled deeply. "In the words of Albert Einstein, *'As far as the laws of mathematics refer to reality, they are not certain, and as far as they are certain, they do not refer to reality.*[6]' Mathematics is a human invention. It is powerful, yes, but not a replacement for experience. I'm afraid, young lady, that you still have much to learn," he said gravely.

Elizabeth's eyes began to burn and she swallowed painfully. In her mind, she'd not only failed to solve the riddle but had also been ridiculed in front of the crowd.

"Acrostic," Lulu whispered in her ear.

"His comments were caustic, weren't they?" Elizabeth replied, holding back tears.

"No, the riddle. It's an *acrostic* poem. Look at the first letter of each line. Lewis Carroll used them a lot," said Lulu. With that, she stood up and announced "Amor. It's Latin for love. Omnia vincit amor – love conquers all!"

"As Lewis Carroll wrote, *'Oh, 'tis love, 'tis love, that makes the world go round!*[7]' The riddle is solved." said the Bandersnatch. Another quick burst of wind sent the papers that had settled on the floor flying about the room once again. Lulu looked toward the window. The Bandersnatch had vanished.

"You did it!" said Elizabeth, a congratulatory smile on her face.

"Council, may we proceed with the Vorpal Blade?" Mrs. Magpie asked. Enough time had been wasted.

The Walrus, who's sagging face looked tired and worn, did not bother to rise from his seat. He merely looked at Mrs. Magpie and nodded his massive head slowly and deliberately.

"Hold out your hand, my dear," Mrs. Magpie motioned to Lulu.

Mrs. Magpie hopped across the table and used her beak to pry open a small drawer. She pulled out a glittery object and dropped it into Lulu's outstretched palm.

"I present you, Lulu, with The Vorpal Blade," she said in a stately manner. The room shook with cheers: "Callooh, Callay! Callooh, Callay!$_1$"

Lulu looked at the tiny sword in her hand, which was only about the length of a pencil. Its handle was inscribed with gorgeous illustrations of dragons and encrusted with pearls (hopefully not from the poor oysters the Walrus and Carpenter had tricked).

"Thank you," she said wholeheartedly. Lulu was surprised (and only slightly disappointed) that the Vorpal Blade resembled a fancy hairpiece more than a formidable weapon. She wanted to ask Mrs. Magpie how such a small blade had slain the fearsome Jabberwocky but didn't want to sound ungrateful.

"The Vorpal Blade will reveal its power when needed," spoke Mrs. Magpie as if she had read her mind.

"Session adjourned," she announced to the crowd. A great sea of fur, scales, and feathers poured out of the room through both doors and windows.

Soon Lulu, Elizabeth, and Mrs. Magpie were the only ones left in the room, which now appeared enormous.

Mrs. Magpie turned to the girls. "Well done, Lulu," she said. "You ladies should return home now. You can use the tree just outside the door. For every ten hours that have passed in Tulgey Wood, only one hour has gone by in your own world. You should be back in time for dinner."

Aha, thought Elizabeth, *that explains Mrs. Magpie's age.*

"Lulu, I expect you back here tomorrow to begin your training. If all goes well during the probationary period, you'll be the official Queen of Tulgey Wood in a fortnight."

Play Along

1. (4 pts.) Using your Tangram set, construct the following puzzle of the Vorpal Blade:

Fun fact: Queen Elizabeth II's coronation regalia included five different swords - three swords representing the values of Mercy, Spiritual Justice, and Temporal Justice, as well as a Sword of State and a Jeweled Sword of Offering. Maybe the Vorpal Blade should be renamed to *The Sword of Math and Magic.* What do you think?

Next, construct this heart to represent "amor":

Fun fact: If you know how to say "love" in Latin ("amor"), you also know how to say it in other languages. It's also "amor" in Spanish, "amour" in French, and "amore" in Italian.

2. (1 pt.) The Bandersnatch spoke of how someone could start with a single fruit and keep splitting it in half. If someone does this, would the fruit ever be completely gone?

3. (1 pt.) To walk to the other side of the room (or anywhere, actually), you have to first walk half of the way. Then you have to walk half of the remaining distance, followed

by half of the distance left after that, and so on. Will you ever get to the other side? Greek philosopher Zeno used this logic to 'prove' that all movement is impossible. Do you think he was correct?

4. (3 pt.) Gather a collection of small objects (all of the same type) like beans, crackers, or blocks. Go through the numbers one by one to determine which ones are prime. Try to arrange that many objects into a rectangle that has more than 1 row. If you can't make such a rectangle, the number is prime.

5. (2 pt.) Continue the Fibonacci sequence: 1, 1, 2, 3, 5, _, _, _, _, _. (remember that each number is made up of the sum of the previous two numbers). Go on a nature walk and count the number of petals on various types of flowers. What do you notice?

6. (4 pt.) Mrs. Magpie said that one hour passes on Earth for every ten hours in Tulgey Wood. If the girls left home at 5:00 pm and spent 15 hours

in Tulgey Wood, what time will it be when they return?

7. (8 pt.) A fortnight is two weeks or 14 days. A day in Tulgey Wood is longer than a day on Earth, lasting 30 hours (or 3 Earth hours). If it is 6:30 pm on Saturday on Earth, then which day of the week will Lulu become Queen?

8. (3 pt.) Write an acrostic poem, where the first letter of each line spells out your name.

11. Two Great Powers

The girls turned to leave when an uncomfortable thought dawned on Lulu. "What about Elizabeth?" she pleaded passionately. "We started this journey together, and she solved just as many puzzles as I. Why can't she be Queen as well? Throughout history, many rulers have shared the throne. I don't mind."

Mrs. Magpie looked down. "That is not within my power, dear," she said sadly.

"Elizabeth has proved herself just as capable as you, but she didn't answer the last riddle correctly. That is the one that counts by the Law."

"Who determines the correct answer to the riddle? Is it all up to that ugly old Bandersnatch?" Lulu retorted. She immediately, clapped her hands over her mouth, realizing that she hadn't spoken in a very queenly manner.

Mrs. Magpie let out a raspy laugh then grew solemn again. "I'm sorry," she said, "but the Law is the Law."

"What if I can prove that both of us were correct?" asked Lulu.

She didn't wait for a response before continuing. "All great forces need balance. Light would not exist without the darkness. Imagination would not have meaning unless it is placed on the backdrop of reality. And I say that love and logic need each other, as well. The more mathematical patterns we uncover in our world, the more reason we have to fill our hearts with awe, reverence, and wonder. Q.E.D.!"

Tiny claps were heard in the back of the room. A Mome Rath, who had stayed behind to survey the floor for food crumbs, had climbed on the back of a chair and was clapping his little hands enthusiastically. Lulu suppressed a laugh.

Mrs. Magpie was unaffected. "Lulu, I'm sorry," she said, "but I cannot change the Law."

Lulu looked at her sister, downhearted.

"Can I say something?" asked Elizabeth quietly, "Nobody ever asked me if I'd like to be Queen. Lulu, do you remember when you wanted to be Alice, but I said that I didn't need to be anyone? Well, I really meant it - I have no grand ambitions to be *somebody*. I'm perfectly happy just being me."

Elizabeth felt better finally saying those words. Since the minute they had been asked to sit on the throne with all those eyes watching them, she had been filled with anxiety. Now all that tension released. She thought of the lines from Emily Dickinson's poem:

How dreary – to be – Somebody!
How public – like a Frog –

To tell one's name – the livelong June –
To an admiring Bog! [7]

Elizabeth didn't dare say the poem aloud and undermine her sister's success, but she felt that it put words to her own feelings perfectly.

"Being Queen is Lulu's dream, and she deserves it," Elizabeth continued, feeling genuinely happy for her sister, "But if it's alright, I would like to attend the training sessions, as I'm always curious to learn new things."

"Can I make Elizabeth the Royal Adviser to The Queen?" Lulu asked.

"If she agrees," said Mrs. Magpie.

"I really don't need a royal title," Elizabeth said, shaking her head. "Isn't a sister like a secret adviser, anyway?" They all laughed.

"It's time for you to go home now," said Mrs. Magpie, "It's been a long day. I'll see you both very soon for your training." She flew out the window in a streak of shiny black feathers.

Lulu and Elizabeth headed outside arm in arm. "Your little speech back there was not technically a mathematical proof, you know," said

Elizabeth as they walked toward the tree. "You probably shouldn't say Q.E.D. at the end unless you've proved something. If you're doing a direct proof, you start by stating your axioms, then-"

"My dear sister," Lulu interrupted, placing one hand on Elizabeth's shoulder and the other on the tree branch, "we have two great powers on our side – love and mathematics. Let's just leave it at that for today, and get ourselves home."

The end.

Jabberwocky

By Lewis Carroll

'Twas brillig, and the slithy toves
Did gyre and gimble in the wabe:
All mimsy were the borogoves,
And the mome raths outgrabe.

"Beware the Jabberwock, my son!
The jaws that bite, the claws that catch!
Beware the Jubjub bird, and shun
The frumious Bandersnatch!"

He took his vorpal sword in hand;
Long time the manxome foe he sought-
So rested he by the Tumtum tree
And stood awhile in thought.

And, as in uffish thought he stood,
The Jabberwock, with eyes of flame,
Came whiffling through the tulgey wood,
And burbled as it came!

One, *two*! One, *two*! And through and through
The *vorpal blade went snicker-snack*!
He *left it dead*, and with *its head*
He *went galumphing back*.

"And hast thou slain the Jabberwock?
Come to my arms, my beamish boy!
O frabjous day! Callooh! Callay!"
He chortled in his joy.

'Twas brillig, and the slithy toves
Did gyre and gimble in the wabe:
All mimsy were the borogoves,
And the mome raths outgrabe.

How Did You Do?

The twins discovered Wonderland,
As Math and Magic led the way.
You were there at every turn,
And even had a chance to play.

Now before we say farewell,
Add up you points so you can see
If you may join Queen Lulu
As mathematical royalty.

Apprentice
Less than 100 points

You've started on your journey
To make yourself a name.
Keep working hard, but don't forget
That math is just a game.

Craftsman
100-149 points

You're getting better at your trade,
Using math to chisel rocks.
Once you stack them into towers,
You will think outside the box.

Aristocrat
150-199 points

You don't need royal blood
Or fancy pedigree
To creatively solve problems
And become math aristocracy.

Noble
200-249 points

Being a lord or lady
Is wonderfully grand.
Keep working on your math skills,
And soon you'll rule the land.

Sovereign
250 points or more

Math Monarch is a title
You've worked quite had to earn.
Now's your turn to use your brain
Toward helping others learn.

About the Author

Lilac Mohr is a software engineer, entrepreneur, and mother of four. She is the developer of the *La La Logic Critical Thinking Curriculum for Preschoolers* (http://www.lalalogic.com), the editor of the poetry anthology *"Classic Poetry for Your Little Genius"*, and the author of the *"Math and Magic"* series of math adventure novels. Lilac holds a B.S. degree in Computer Information Systems and an M.S. degree in Statistics.

Lilac's Blog:
http://learnersinbloom.blogspot.com

On Facebook:
http://www.facebook.com/learnersinbloom

If you enjoyed this book I wrote,
(If it succeeded in floating your boat),
Please leave a review-
It's the least you can do
To cast your Amazon "vote".

Appendix A: Solutions

Chapter 1

1. To use alliteration, come up with words that start with the same letter.

2. Rajveer Meena from India recently recalled 70,000 digits of Pi, which took him ten hours. How did you do?

3. "I prefer Pi." (said by Lulu) and "Was it a cat I saw?" (said by Elizabeth) are both palindromes. Can you make up your own palindrome sentence?

4a. There are 9 two-digit palindromes (11-99).

4b. There are 90 three-digit palindromes. The first (and last digit) can be from 1 to 9. For each of those nine options, there are 10 possible values for the middle digit. 9 x 10 = 90.

5a. On Mercury, Mrs. Magpie is 146 x 4 = 584 years old.

5b. On Mars, Mrs. Magpie is 146 divided by 2, which equals 73.

Chapter 2

1. If you had trouble making all the Tangram shapes, search online for a Tangram template to print and cut out.

2. Tangram bird:

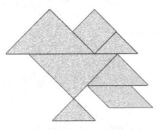

3. Using Gauss's method, you can figure out the sum of the numbers from 1 to

16. There will be 8 pairs of numbers that each add up to 17. 17 x 8 = 136. There are 4 rows in the magic square which will all have the same sum. If you divide 136 by 4, you will get 34. Each row, column, and diagonal will total 34.

4. The solution to the 4 by 4 magic square:

8	11	14	1
13	2	7	12
3	16	9	6
10	5	4	15

5. There are 9 squares made up of 4 cells each.

6. What makes this magic square extra special is that the sum of the four numbers making up the nine smaller squares all add up to 34, as well.

Chapter 3

1. Tangram key:

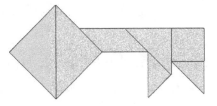

2. EGGHEAD

3. There are 6 possible letters for the first position, 3 vowels for the second, and another 6 for the third, so:

6 x 3 x 6 = 108 permutations.

4. Lulu is correct. An even number becomes odd when one is added or subtracted. An odd number becomes even when one is added or subtracted from it.

5. a. even + even = even
   b. odd + odd = even

c. even + odd = odd

6. The symbols are as follows:

Chapter 4

1. Tangram rabbit:

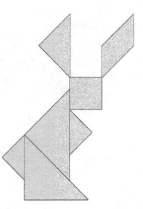

2. Were you able to draw a freehand circle? Some people use their hands or arms as compasses to accomplish this task.

3. Did your compass help you draw a perfect circle? The compass is a symbol of precision. You can find it in the logo of the Freemason organization, which began as a craft guild for stonemasons. A stonemason shapes rocks into precise shapes for use in building structures.

4. How close did your measurements come to Pi?

5. a. Descartes
   b. Galileo
   c. Pascal
   d. Newton

<u>Chapter 5</u>

1. Tangram Mobius strip:

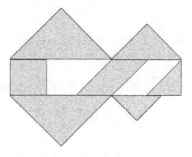

2a. A regular loop has two sides - an inside and an outside. You could color these sides in two different colors that will never touch.

2b. A Mobius strip only has a single side, and can only be colored with one color.

3. When you cut the regular loop in half, you get two thinner loops. When you cut a Mobius strip in half, it stays a Mobius strip, but becomes thinner. Here's the complete limerick:

A mathematician confided,
That a Mobius band is one sided,
And you'll get quite a laugh
If you cut one in half,
For it stays in one piece when divided.

4. When you cut a Mobius strip in thirds, you get a loop attached to a thinner Mobius strip.

5. Your fractal tree might look similar to this one (which was generated by a computer):

6. How does your tessellation look? Maybe you can be the next M.C. Escher. You can use the same process to make tessellations out of other shapes, as well, as long as they tile. Hexagons tile

with each other, or you can use a combination of shapes like octagons and squares together.

Chapter 6

1. Tangram boat:

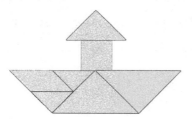

2. Here are the trips that were made and the amount of water in the boat:

Start: 1 cm
L+P- New: 1.5, Total: 2.5
L: New: 0.5, Total: 3.0
M+L - New: 1.0, Total: 4.0
P+L - New 1.5, Total 5.5
E+L- New 1.0, Total 6.5
L- New 0.5, Total 7.0
P+L- New: 1.5, Total: 8.5

By the end of the river crossing, there was 8.5cm of water in the bottom of the boat.

3. First, two of the girls go across, and one comes back. Then an adult goes across the river while the girl already across takes the boat back. The process is then repeated to get the remaining adult across the river (two children, one comes back, the adult goes across). The girl on the other side then brings the boat back to pick up her sister.

Chapter 7

1. Tangram bridge:

2. Did you learn some interesting things about Lewis Carroll? Here are the origins of the places on Lulu and Elizabeth's map:

Lake Charles and Dodgson: Lewis Carroll's real name was Charles Dodgson.

Camden-Town: He wrote a poem titled "Atalanta in Camden-Town".

Chataway Isle: Gertrude Chataway was a young girl who inspired Carroll's poem, *"The Hunting of the Snark"*. In the poem's dedication, Carroll included an acrostic poem, whose first letters spell "Gertrude Chataway".

Liddell Island: Alice Liddell was the "real Alice", from *"Alice in Wonderland"* and *"Through the Looking Glass"*. Carroll wrote an acrostic poem for Alice Liddell in his book, as well.

Did you learn in your research that Lewis Carroll was not only an author and poet but also a mathematician?

3. Welcome to graph theory! Your model may look something like this:

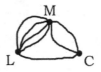

4. Add another bridge between Liddell and Chataway Islands:

5. The easiest way to solve these problems is to work backward. Start with 1, and double the number for each bridge.

A. For 3 bridges, you need 8 fruits.
B. For 5 bridges, you need 32 fruits.
C. For 9 bridges, you need 512 fruits.
D. For 15 bridges, you need 32,768 fruits!

Since each time, the number of fruit doubles, we can take the number of bridges and multiply 2 by 2 that number of times. For example, for 3 bridges, you need 2 x 2 x 2 = 8 fruit. For 5 bridges, you need 2 x 2 x 2 x 2 x 2 = 32 fruit. It's good arithmetic practice to compute this by hand, but you can also use a calculator. Look for a button that looks like this: $\boxed{x^y}$ It's called the exponent button. Ask an adult to help you if you can't find a button like this on your calculator since it might have a different label. See if you can figure out this exponent button works. Try entering 2, then the exponent button, then 5, and see what you get. Maybe this will come in useful for Lulu and Elizabeth as their adventure continues (hint, hint).

6. WISH, WISE, RISE, RICE, RACE, RATE

7a. One possible solution:
CAT, COT, DOT, DOG

7b. One possible solution:
FOOD, GOOD, GOLD

7c. One possible solution:
SEED, FEED, FLED, FLEE, TREE

chapter 8

1. Tangram kangaroo:

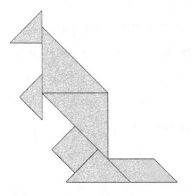

2. It would always say "Yes, I'm a Mome Rath."
The Mome Rath would be telling the truth, and
the Slithy Tove would be lying.

3. Alex is a Slithy Tove, and Betty is a Mome
Rath. If Alex were telling the truth, then Betty
would agree with him, so he has to be lying.

Only a truth-teller would say that a liar is not to be trusted, so Betty must be a Mome Rath.

4. Bob and Carl are both Slithy Toves. Ann is a Mome Rath. Nobody would ever claim to be a Slithy Tove, so Bob was lying when he claimed that Ann said that. Similarly, Carl would never say that he was a Slithy Tove, so he must be lying about Ann being a Slithy Tove.

5. Did you discover, like Archimedes, that the amount of water that is displaced by an object depends on its volume (the space it takes up) rather than its weight?

6. You only have to weigh the coins once. Put one coin on each side of the balance scale and set one aside.

7. You can figure it out in two weighings or less. Place one coin on each side of the balance scale and set one aside. If the coins on the scale balance, then you're done (the counterfeit is the coin you set aside). If they do not balance, you

need to weigh one of the coins against the coin that was set to the side.

8. You can only guarantee that you'll find the answer in 4 weighings. The first weighing has three coins on each side of the balance scale, and three set aside. In the worst case scenario, the scale does not balance. You've eliminated the case where all the coins are the same weight, and you've narrowed the culprit down to 6 suspects.

For the second weighing, remove one coin from each side of the scale and put those two coins aside. If the scale still does not balance, you've narrowed it down to 4 suspects.

For the third weighing, remove one more coin from each side (leaving one coin on each side of the scale, and two extra coins on the table). Whether the scale balances or not, you've narrowed it down to two suspects (either the two coins on the scale or the two that are not on the scale).

For your fourth weighing, compare one of the 'suspects' with a coin that you know is not counterfeit. If it does not balance, then the 'suspect' you are weighing is the counterfeit. If it does balance, then the other 'suspect' is the counterfeit. It looks like Elizabeth was correct, after all!

chapter 9

1. Tangram candle:

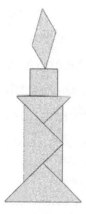

2. Your chart should look like this:

Day	Number	Sum
1	1	1
2	2	3
3	4	7
4	8	15
5	16	31
6	32	63
7	64	127

Did you notice any patterns? On Day 2, the number owed is 2. On Day 3, the number owed is 2 x 2 = 4 (2 to the power of 2). On Day 4, the number owed is 2 x 2 x 2 = 8 (2 to the power of 3). So to calculate the amount owed on a certain day, subtract 1 from the number and multiply 2 by itself that many times. For example, on day 7, the amount owed is "2 to the power of 6", which equals 64.

The sum is equal to the number owed on the next day minus one, so all you have to do to calculate the total owed by day "n" is to take "2 to the power of n" and subtract 1. For day 7, for example, take "2 to the power of 7" and subtract 1 to get 127 for the running total.

3. Were you able to duplicate Mrs. Magpie's "pecks" on your own calculator?

4. Use your calculator to compute "2 to the power of 15" to get 32,768 cents. Subtract 1, and then move your decimal point to get $327.67. This amount is a lot closer to the $200 reward that Mimsy Kangaroo initially proposed. What a difference five days makes when it comes to exponents!

5. You should notice that the number of moves looks just like the "sum" column in the chart you made in question 2. For "n" disks, the number of moves is "2 to the power of n, minus 1".

6. "2 to the power of 64, minus 1" is 18,446,744,073,709,551,615 days. That's over 500 billion years!

chapter 10

1. Tangram sword:

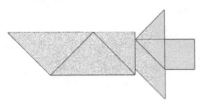

Tangram heart:

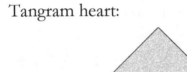

2. A mathematician would tell you that you can keep cutting the fruit in half forever. In practice, it is not so easy to accomplish. With an adult's permission, take a fruit (like a grape) or even a piece of clay and see how many times you can manage to cut it in half

244

3. Philosopher Zeno cannot be correct because the motion is possible (you can reach the other side of the room). When you get into more advanced math, you'll be able to prove that 1/2 + 1/4 + 1/8 + 1/16 + 1/32... = 1.

4. The prime numbers less than 40 are 2, 3, 5, 7, 11, 13, 17, 19, 23, 29, 31, and 37. How many did you find?

5. 1,1,2,3,5,8,13,21,34,55. You'll find the number of petals of many types of flowers (especially those in the Daisy family) in the Fibonacci sequence.

6. 15 hours in Tulgey Wood are the same as one and a half hours on Earth. If they left at 5:00 pm, they will return at 6:30 pm.

7. 14 days x 3 hours = 42 hours (on Earth). To add 42 hours to 6:30 pm on Saturday, you can add the time in pieces. 24 hours (one Earth day) later would be 6:30 pm Sunday. 12 more hours to get to 6:30 am on Monday. 6 hours later

would be 12:30 pm on Monday. That is when Lulu will become Queen.

8. If you found that creating an acrostic poem was easy, try creating a poem that is a double acrostic. In a double acrostic, one word is spelled using the first letter of each line, and another letter is spelled by the last letters of the lines.

Appendix B: References

1. Lewis Carroll, <u>Alice's Adventures in Wonderland and Through the Looking Glass</u> (New York: Penguin, 2010).

2. W. H. Davies, <u>Songs Of Joy and Others</u> (London: A. C. Fifield, 1911).

3. William Shakespeare, <u>Romeo and Juliet</u> (London: Arden Shakespeare, 1980).

4. J. A. Lohne, "Isaac Newton: the rise of a scientist, 1661-1671" <u>Notes and records of the Royal Society, vol 20</u> (1965).

5. D. E. Smith, <u>A Source Book in Mathematics</u> (Cambridge, MA Harvard University Press, 1969).

6. Albert Einstein, <u>Sidelights on Relativity</u> (New York: Dutton & Co., 1923).

7. R. W. Franklin, ed. <u>The Poems of Emily Dickinson: Reading Edition.</u> (Cambridge, MA: The Belknap Press, 1999).

Appendix C: Lulu and Elizabeth's Bookshelf

1. E. Nesbit, <u>The Book of Dragons</u> (San Francisco: Chronicle, 2007)

2. Molly Perham, <u>King Arthur and the Legends of Camelot</u> (New York: Viking, 1993)

3. Hugh Lupton, <u>The Adventures of Odysseus</u> (Cambridge, MA: Barefoot Books, 2006)

4. Louisa May Alcott, <u>Little Women</u> (London: Puffin, 1994)

5. Robert Louis Stevenson; Scott McKowen <u>Treasure Island</u> (New York: Sterling Pub, 2004)

6. James Baldwin, <u>Don Quixote for Young People</u> (New York, Cincinnati, Chicago: American Book Company, 1910)

7. Luetta Reimer; Wilbert Reimer, <u>Mathematicians Are People, Too: Stories From the Lives of Great Mathematicians</u> (Palo Alto, CA: Dale Seymour Publications, 1990).